T0178302

SpringerBriefs in Energy

More information about this series at http://www.springer.com/series/8903

Ralph G. Scurlock

Stratification, Rollover and Handling of LNG, LPG and Other Cryogenic Liquid Mixtures

 Springer

Ralph G. Scurlock
University of Southampton
Southampton, UK

ISSN 2191-5520 ISSN 2191-5539 (electronic)
SpringerBriefs in Energy
ISBN 978-3-319-20695-0 ISBN 978-3-319-20696-7 (eBook)
DOI 10.1007/978-3-319-20696-7

Library of Congress Control Number: 2015951092

Springer Cham Heidelberg New York Dordrecht London

Springer International Publishing AG Switzerland is part of Springer Science+Business Media (www.springer.com)

Preface

My interest and experience in cryogenic engineering began in the early 1960s when we set up the first postgraduate training course on "Cryogenics and Its Applications". At that time, cryogenics meant LOX, LIN, LH2 and LHe. This course was developed to involve industry in the teaching and the hands-on demonstrations, together with a 3-month project in industry for all the students in participating companies.

Very soon, we found the project work was taking us into 3-year Ph.D. research programmes on a wide range of interdisciplinary topics, together with expanding into higher temperatures within the world of hydrocarbons, using a purpose-build LNG Safety Laboratory. This expansion was also taking us into the new territory of irreversible thermodynamics and the extraordinary properties of cryogenic liquid mixtures. LNG was one of these mixtures, but not the only one or the first one we met.

At that time (1960), LPGs with normal boiling points between −40 and −2 °C were being developed as an industrial fuel by liquefying the gas being flared off at refineries and oil wells. We were soon introduced to the problems of handling LPGs in bulk and surprised to find that our experience with colder cryogenic liquids such as LOX, LIN and LHe enabled us to begin solving the LPG problems very quickly.

At the same time, we began to find that liquid mixtures had unique properties of their own, which needed research programmes to gain an understanding of these properties, outside our experience with single component cryogenic liquids.

So, when the LNG rollover event happened at La Spezia in 1971, we were quickly involved in trying to understand what had happened. Subsequent research programmes, using the more sophisticated flow visualisation and measurement techniques we developed in the later 1970s, enabled us to reproduce and study rollover events between stratified layers of cryogenic liquid mixtures. In particular, we were able to visualise what happens in a rollover event using laser Doppler and photographic and video systems.

Continuing contact with the LNG industry, via contributions to the LNG cargo management courses at the Warsash Maritime School of Navigation, Southampton, led to 3- and 12-month Institute of Cryogenics training courses on LNG technology.

These courses at Southampton and Sonatrach Gas School, Boumerdes, Algeria, continued from the mid-1970s for many years.

Continuing research on cryogenic mixtures provided a firm base for developing the general features of cryogenic fluid dynamics as a new discipline for all cryogenic systems. As a consequence, a recommendation was made to redefine "cryogenics" with a broader base to cover all liquids boiling below 0 °C, the ice point.

This recommendation has not been accepted by everyone, but will be used for the purpose of this text, as a basis for learning to understand LNG and LPG rollovers. Our research findings have been published in some 150 papers scattered throughout various cryogenic and technical journals, cryogenic engineering conference proceedings and textbooks. Most of the material in this book is based on these findings, much of which comes from M.Sc. and Ph.D. theses of many of my students at Southampton University.

Southampton, UK   Ralph G. Scurlock

Acknowledgements

My grateful acknowledgements include, in particular, the following former Ph.D. research students in chronological order: P. Lynam, A. Mustafa, M. Wray, W. Proctor, J. Boardman, D. Richards, G. Beresford, O. San Roman, A. Tchikou. R. Rebiai, M. Atkinson-Barr, Y. Wu, S. Mirza, M. Wu, T. Agbabi, S. Yun, J. Shi and A. Thomas.

I am greatly indebted to the teaching staff and research fellows of the Institute of Cryogenics, University of Southampton, Prof. C. Beduz, Prof. J. Watson, Prof. Y. Yang, Dr. P. McDonald, Dr. M. Islam, Dr. G. Rao, Dr. A. Tavener and Dr. M. Burton.

I am also much indebted to many university staff who contributed to and collaborated with the development of cryogenic fluid dynamics and its many applications, thereby broadening each other's cryogenic research and training activities, including, Prof. G. Lilley and Prof. M. Goodyer, Dept. of Aeronautics and Astronautics (fluids dynamics, cryogenic wind tunnels, space applications); Prof. R. Bell, Prof. R. Farrar, Prof. S. Hutton, Dr. D. Wigley and Mr R. Bowen, Dept. of Mechanical Engineering (insulations and materials); Prof. G. Hills and Dr. A. Rest, Dept. of Chemistry (low-temperature chemistry, FTIR spectroscopy); Dr. R. Craine, Dept. of Mathematics (CFD modelling of convective mixing and rollover); and Capt. G. Angas, School of Navigation, Warsash (handling problems with LNG and LPG sea tanker cargoes).

I am much indebted to Chris Clucas, an M.Sc. cryogenics graduate of Southampton University, now with Bernhard Schulte Ship Management Group and inaugural president of the Society for Gas as a Marine Fuel, for suggesting I should write this monograph on rollover and for reviewing my manuscript.

Finally, I am extremely grateful to my wife Maureen for typing and correcting the many drafts of this text.

Contents

List of Figures

Chapter 1
Advisory Summary and Introduction to LNG (and LPG)

Abstract If two non-boiling liquids at ambient temperatures are mixed (say diesel and gasoline fuels) the heat of mixing causes the temperature of the mixture to rise; that's all.

If two LNG liquids with different compositions, or LPG liquids, are mixed at atmospheric pressure, the heat of mixing will create a large volume of boil-off vapour.

This chapter introduces this and other differences between ambient temperature and cryogenic temperature liquid mixtures, which need to be understood by all users.

The chapter begins with a summary of 15 points towards efficient storage, handling and use, together with some recommendations to prevent stratification and unstable evaporation behaviour, including roll-over.

The chapter concludes with a list of definitions of terms used throughout this monograph.

1.1 Advisory Summary on the Storage and Handling of LNG (and LPG) Mixtures

The current boom in world-wide trade in LNG, and the growing breadth of industrial and domestic applications requires a basic understanding for dealing with these cryogenic liquid mixtures of Liquefied Natural Gases or LNG. This monograph commences with an advisory summary on how to handle LNG, followed by several chapters describing the basics; and then the main chapters discuss the actions to be taken in the storage and handling of LNG (and Liquefied Petroleum Gases or LPG where appropriate).

1. LNGs are colourless, multi-component liquid mixtures of methane (density 422 kg/m³), ethane (density 544 kg/m³) and smaller proportions of propane (density 581 kg/m³), higher hydrocarbons, carbon dioxide and nitrogen (density 807 kg/m³), with normal boiling point at 1 bar in the range 112–120 K (or −161 to −153 °C).

LPGs are colourless, multi-component liquid mixtures of propane (density 581 kg/m^3), n-butane (density 601 kg/m^3), iso-butane (density 594 kg/m^3) and small proportions of higher hydrocarbons, with normal boiling point at 1 bar in the range 231–272 K (or −42 to −1 °C).

2. The proportions of the LNG components vary widely, depending on the source or gas field. Since the pure components have widely different densities, the density of a particular LNG can vary with its composition, in addition to having a strong density variation with temperature. The net result is a complex liquid with storage problems associated with internal density differences, particularly on a large scale.

3. The boiling liquid surface is covered by a froth of bubbles, while the container walls are wet with rising films of ethane rich liquid. With its low viscosity and surface tension, the liquid is easily disturbed and sloshes readily when its container is moved, accompanied by crackling sounds and spitting of LNG droplets up to the container roof, from local vapour explosions in the frothing surface.

4. There are two types of storage (1) large scale (up to 300,000 m^3 capacity) tanks, with gas purged insulations storing LNGs at 1 bar pressure for shore, ship and floating systems, and (2) relatively small scale VI pressure vessels (up to 1000 m^3 capacity) with vacuum insulation, storing LNGs at pressures up to 24 bar for widespread use as a cryogenic fuel for road and rail vehicle propulsion, as a ship bunkering fuel and for local industrial storage. (LPGs are usually stored and distributed for widespread use in small scale, pressure vessels, under pressure at ambient temperatures, with no thermal insulation. For large scale storage, LPGs are stored as cryogenic liquids at 1 bar in insulated tanks as described in this monograph when appropriate.)

5. Liquefaction of LNG yields a 500–620 fold reduction in volume at 1 bar for bulk transport and storage, as well as for industrial and domestic energy usage. (Liquefaction of LPG yields a 240–310 fold reduction in volume at 1 bar.)

6. The liquefied gases, with which this monograph is concerned, have normal boiling points (at 1 bar pressure) well below ambient temperatures. Storage therefore requires thermal insulation to reduce evaporation rates (or boil-off rates, BOR) to minimum practical values.

7. For safe storage at 1 bar pressure. The total heat flow entering the stored liquid should be balanced by a normal boil-off rate, BOR, such that

$$\text{Heat flow IN} = \text{Latent heat of evaporation} \times \text{BOR.}$$

8. If the BOR drops below the average normal figure, then some of the heat inflow is being stored in the liquid as thermal OVERFILL. The subsequent uncontrolled release of this stored energy by sudden, increased evaporation above the normal BOR, is called a ROLLOVER.

9. Density equilibration between two stratified layers, by mechanical mixing, will yield significant evaporation from the heat of mixing of the two compositions, as well as from energy release of thermal overfill.

10. If stratification is detected or auto-stratification is suspected, then urgent action is needed to mix the contents of each tank. Allowing the stratification to continue will allow more thermal overfill to build up in the lower layer, and increase the amount of vapour produced subsequently by a mixing operation or a rollover.

11. The main problem with a rollover is the considerable and uncontrolled amount of vapour generated by the spontaneous mixing over a short period of time. The BOR is unpredictable and may rise slowly by 10–20 fold, or rise rapidly in seconds to a peak of 100–200 fold All vents to the atmosphere must open to release the vapour and reduce the rising pressure.

12. The vented LNG vapour will generate a cold, low level, white vapour cloud drifting downwind, which must not reach any ignition source. As the cold vapour cloud mixes with atmospheric air, the water vapour freezes to form the observed white cloud. As mixing with air continues, the temperature of the natural gas vapour rises and the white condensate evaporates.

    Eventually the density of the vapour cloud becomes less than the air density and the cloud rises rapidly above the ground or sea, clear of possible ignition sources.

    However, for LPG vapour clouds, the cold, low level, vapour cloud density remains heavier than air (by up to 30–40 %), as it drifts down-wind and mixes with atmospheric air. As its temperature rises and the white condensate evaporates, the cloud becomes invisible but still remains close to the ground (or sea) and will drift unseen down-wind for several kilometers. During this time it must not encounter any ignition source.

13. If LNG is spilled onto water, it will evaporate to the accompaniment of bangs and crackles from large numbers of small vapour explosions. This behavior is alarming but not dangerous. The explosions do not appear to generate enough energy to spark ignition of the evaporating LNG pool.

14. Double walled, vacuum insulated, VI tanks have internal vessels capable of storing LNG with zero boil-off under pressure up to 24 bar, and appear to have little likelihood of spontaneous mixing or rollover. These have relatively small volumes of from 0.1 to 1000 m^3, for use as fuel tanks for LNG.

15. The larger storage tanks with gas-purged insulation, and with liquid storage volumes up to 300,000 m^3, operate with boil-off at 1 bar, fractionally greater than atmospheric pressure. Unlike the smaller VI tanks, it is these larger tanks which are sensitive to the consequences of density stratification, rollover and other boil-off instabilities.

1.2 Introduction: 1969 Warning Against Stratification and "Rollover" on LPG Tankers

In the 1960s, it was common practice to flare off petroleum gases at oil wells and oil refineries, while Liquefied Petroleum Gases (LPG) were a relatively small global business using refrigerated tankers of 5000–10,000 m^3 capacity to collect LPGs of various compositions from a number of oil refineries.

During an early consultation on LPG and its properties, I learned that LPG part cargoes were being loaded layer by layer in the same ship tank. Mixing operations were not allowed in port, and were started only when the tanker left port and entered the open sea.

I advised against this unsafe practice, of loading different cargoes of LPG within the same tank without mixing. The heats of mixing for LPGs were large and could generate an uncontrolled boil-off of LPG vapour, i.e. a rollover (the term "Rollover" was not used until 1972 after the La Spezia LNG incident). The practice was very quickly stopped in 1969, communicating by Telex to many but not all LPG tankers.

Since then, the formation of SIGTTO (the Society of International Gas Tanker & Terminal Operators) and subsequent IMO Marpol regulations restrict on-board mixing operations in all tankers; it is no longer permitted during a sea voyage- but can be carried out at anchorage, or at the berth.

The operation of mixing LPGs on-board involves the recondensation of the considerable vapour generated and returning the condensate to the same or a different tank. The rate of mixing is set by the available refrigeration on board and the rate at which vapour is generated.

However, it was noticed that when a cold propane-rich liquid was added into a warmer butane-rich heel, the amount of vapour generated was twice as great as that produced when the hotter butane was added to a colder propane-rich heel. The first type of mixing and recondensation by on-board refrigeration therefore took twice as long, and was therefore commercially unattractive.

I was asked to explain this phenomenon as my second contact with the LPG industry. The results of our experimental studies were that, being thermodynamically irreversible, mixing was indeed path-dependent, and we were able to confirm that adding hot to cold yielded twice as much vapour as adding cold to hot. This was no artefact and was indeed an important indication on how to mix all cryogenic liquids efficiently.

These mixing phenomena also emphasised that the fluid properties of hydrocarbon liquid mixtures with multi-component compositions were much more complicated than those of single component cryogenic liquids, with some unexpected behaviours which affected the safe storage and handling in large containers and tanks.

However, contact between LPG operations and the newly developing LNG business was non-existent. Hence the 1969 warning against stratification in LPG, and the immediate actions taken to stop it in LPG sea tankers went un-noticed in the separate business of handling LNG. As a result, the scene was set for the first LNG rollover incident 2 years later in August 1971 at La Spezia, Italy, and possibly the unreported LPG rollover in a giant 100 m diameter tank in Dubai in the early 1980s.

1.3 Discovery of the Unstable Evaporation of All Cryogenic Liquids

For many years, we had come to accept that studying the evaporation behaviour of cryogenic liquids yielded some unexplained and unexpected characteristics.

This acceptance stopped abruptly in 1975, when we were approached by the Atomic Energy Research Establishment, Harwell, UK, with a problem concerning explosions with 50 L LIN storage vessels. The Harwell committee of enquiry had not found any mechanical reason for the dewar failures. The dewars had been cold and had contained LIN at the time of each event. High pressures had developed in the dewars, causing them to explode and creating damage to surrounding equipment, together with parts flying through the roof.

It was not at all clear to us what had caused the explosions. Nevertheless, AERE, Harwell needed an answer and suggested we set up some monitoring tests of their large storage dewars, which we did after, after putting steel screens around the test dewars.

At that time, there was no instrument on the market, and the first step was to build an electrically recording, instantaneous gas-flow meter. The first boil-off readings were plagued (by DC noise, or so we thought) by large, random, rapid variations in flow rate with a time constant in the range of 1–10 s. After a month or so, rebuilding and checking the home-made electronics, we came to realise that the rapid variations, by up to 10 % of the average boil-off, reflected the normal irregular evaporation of a cryogenic liquid.

Occasionally, much larger flow rate spikes were observed. Tapping or shaking a dewar of LIN also induced large flow rate spikes. These spikes were the first indication of what are now called "vapour explosions", when the rate of vapour production could exceed the limiting, choked flow capability of the vent to the atmosphere or the narrow 20 mm necks of the dewars in use. The solution to the problem was therefore to use 50 mm wide-necked dewars, so that choked flow was never reached under the vapour explosions being experienced.

At the time of the 1975 AERE, Harwell studies, we were raising more questions than answers.

It was clear we did not understand at all the boil-off behaviour of single- and multi-component cryogenic liquids in general, or of LIN and LNG in particular. A programme of research was therefore set up, with master and doctoral students taking part under the supervision of post-doctoral research assistants and with the availability of a specially constructed LNG safe-handling laboratory.

With financial support from the Science Research Council and industry, this work resulted in a clear understanding of cryogenic liquid behaviour over the next 20 years. In particular, it included the study of some 100 fully instrumented rollover events during simulation experiments of mixing between two cryogenic liquid layers of different density and composition [1].

1.4   The Contents of This Monograph

The monograph begins with a summary of the storage and handling behaviour of the cryogenic liquid mixtures of LNG and LPG, together with recommendations on how to handle and prevent unstable evaporation phenomena, including rollovers [2, 3]. Two contexts are covered to meet present day applications, namely (a) large 1 bar tanks on shore, and afloat on platforms and in sea-going tankers. (b) Relatively small pressure vessels up to 3–4 m diameter for LNG fuel applications.

This monograph on cryogenic hydrocarbon liquid mixtures is based on experimental results obtained over the past 50 years or so, and contains practical interpretations of the explanations to be made of their unstable evaporation phenomena, based on cryogenic fluid dynamics.

The monograph first looks at the various sources of heat flow into a cryogenic system, and considers how these may be reduced to acceptable low values by appropriate insulation techniques. The monograph then develops an understanding of the surface evaporation process and how it can break down under convective instabilities.

After these basic chapters, the monograph develops an understanding of rollover, and how this phenomenon is a common property of all cryogenic liquid mixtures in constant pressure tanks with boil-off at 1 bar. Stratification is the prerequisite of rollover, and options for removing stratification by effective mixing are discussed. The sequence of a rollover event is then discussed, to provide a basis for cargo management decisions.

A chapter is included on storing and handling LNG in vacuum insulated VI pressure tanks with zero loss of content and no rollover.

A chapter on liquid handling then sets out the need to use sub-cooled liquid, to avoid 2-phase flow for trouble-free transfer.

A final chapter on safety includes some personal comments on working with LNGs and LPGs.

Before starting with the basic chapters, some definitions of the terminology used in this monograph are introduced (see Table 1.1). These definitions may not coincide with those used by steam engineers, so be warned!!

Table 1.1 Basic properties of cryogens at T_S (NBP)

	MW	T_M (K)	T_S (K)	T_C (K)	P_C (bar)	ρ_L (kg/m³)	λ (kJ/kg)	V ratio[a]	ρ gas[b]
Nitrogen	28	63.15	77.31	126.3	33.99	806.8	198.9	680	0.98
Methane	16	90.69	111.7	190.5	46.0	422.4	510.3	622	0.56
Ethylene	28	104.0	169.4	282.3	50.4	568	482.6	478	0.97
Ethane	30	90.35	184.6	305.3	48.71	544.1	488.5	427	1.05
Carbon dioxide	44	–	194.7[c]	304.3	73.0	1560[c]	563.0[c]	823[c]	1.52
Propane	44	85.45	231.1	370.0	42.1	581	426	312	1.52
n-Butane	58	138	272.6	426	36.0	601	386	241	2.00
Water	18	273	373	647	217.7	1000[d]	2257	1800	0.62

[a]V ratio = volume ratio gas/liquid at 288 K
[b]ρ gas = gas density relative to air at 288 K
[c]Sublimation values
[d]At 373 K

1.5 Definitions of Liquid States

1.5.1 The Definition of a Cryogenic Liquid, with Normal Boiling Point Below 273 K

In this monograph, time after time you will see that if a liquid is at a temperature below ambient, it is subject to heat inflows and associated natural convection phenomena, which together determine its storage behaviour and handling properties.

There are no other differences in basic physical properties between ambient temperature liquids, like water or gasoline, with normal boiling points above ambient, and hydrocarbon liquids with normal boiling points below ambient.

Therefore, for the purpose of this monograph, we have adopted the definition of a cryogenic liquid as one with a normal boiling point below the triple point of water at 0.13 °C, or 273 K. The application of cryogenic fluid dynamics is then developed to describe and predict the behaviour of LNG, LPG and other hydrocarbon mixtures with NBPs below 273 K [4, 5].

Figure 1.1 shows the general equilibrium or saturation P–T curve, together with the various states to be defined below.

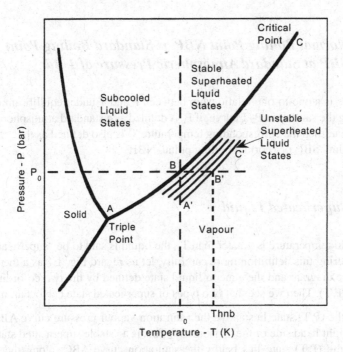

Fig. 1.1 P–T diagram for a typical cryogenic liquid (T_{hnb} = homogeneous nucleate boiling temperature)

1.5.2   Boiling Temperature

Under atmospheric pressure P_0, the "boiling temperature" of the liquid T_0 is the temperature at which the freely boiling liquid cryogen is in equilibrium with its vapour at pressure P_0.

1.5.3   Saturation Temperature and Saturation Vapour Pressure

The boiling temperature T may also be called the "saturation temperature", while P may be called the "saturation vapour pressure".

1.5.4   Saturation Vapour Pressure Curve

The saturation vapour curve along ABC from triple point to the critical point is the accepted relationship between P and T for a single component liquid.

1.5.5   Normal Boiling Point NBP or Standard Boiling Point SBP at Standard Atmospheric Pressure of 1 Bar

Since there is a one-to-one relationship between P and T under equilibrium conditions along the saturation P–T curve, if P_0 is defined as a standard atmospheric pressure of 1 bar, then the corresponding temperature T_0 is also defined as the "standard boiling point" SBP, or "normal boiling point" NBP.

1.5.6   Superheated Liquid

If the liquid temperature is greater than T_0, the liquid is said to be "superheated".
    Considering this definition more carefully, let us regard Fig. 1.1 as a thermodynamic state diagram and the general liquid state defined by the two co-ordinates P and T, or (P, T). Then we see that two types of superheated state can occur, namely:

1. When the (P, T) state lies above the saturation vapour pressure curve ABC, and to the right hand side of the T_0 isotherm. This is a "stable superheated state".
2. When the (P, T) state lies below the saturation curve ABC, along the curves A′B′C′. This is the "unstable superheated state" from which the liquid normally evaporates or changes state.

It will be shown later that the bulk liquid is superheated above T_0 by a finite temperature difference necessary to drive the surface evaporation process, from a fraction of 1 K up to as high as 8 K. This superheated state of bulk liquid in storage is the unstable type, characterised by the curves $A'B'C'$.

A superheated liquid state is generated when heat is added to the liquid commencing at T_0 but without change of phase, and with or without an increase in pressure.

1.5.7   Liquid Superheat

"Liquid superheat" is a loose term for the temperature difference $(T-T_0)$ above T_0, or the enthalpy increase $(H-H_0)$ associated with the superheated liquid.

1.5.8   Thermal Overfill

Thermal overfill is the additional heat energy taken up by a superheated liquid and is also equal to $(H-H_o)$.

1.5.9   Thermal and Pressure Subcooled Liquids

If the liquid state (P, T) lies above the saturation curve ABC and to the left hand side of the T_0 isotherm, then the liquid is said to be "subcooled", or strictly, "thermally subcooled". The term subcooled also applies if the liquid temperature remains at T_0 while the pressure is raised above P_0 i.e. the liquid is pressure subcooled.

1. A thermal subcooled liquid state is generated when heat is removed from the liquid, initially at T_0, and the degree of subcooling can be specified by the enthalpy removed (H_0-H).
2. A pressure subcooled liquid state is generated when the vapour space pressure is increased above P_0 at constant T_0. There is little change in enthalpy and the degree of pressure subcooling may be specified in terms of the pressure head $(P-P_0)$.

1.5.10   Wall Superheat

Under heat transfer between solid and fluid, the solid surface must be at a higher temperature than the fluid in contact with, or immediately adjacent to it, to drive the heat into the fluid. The temperature difference to facilitate the heat transfer is called the "wall superheat".

For convective heat transfer to a liquid, the wall superheat is, for example, from 0.01 to 0.1 K in magnitude for LNG, driving low heat fluxes into a liquid boundary layer flow. For convective heat transfer into a vapour, the wall superheat is, for example, of the order of 0.1–1.0 K for cold methane vapour.

For nucleate boiling heat transfer, the wall superheat is from around 1 up to 10 K, when the wall heat fluxes can be very large.

1.5.11   Boil-Off and Boil-Off Rate, BOR

The terms "boil-off" and "boil-off rate, BOR" are strictly applicable only when the liquid is boiling by the heat transfer process of nucleate boiling. In the majority of storage situations, there is only evaporation from the surface of the liquid and there is no boiling. The term "evaporation rate" is then the correct terminology to use. However, "boil-off" is in common use to describe all liquid evaporation and is frequently used in this monograph.

1.5.12   Heat Flux and Heat Flow

The term "heat flux" is the quantitative wording for heat flow per unit area and is usually measured in kW/m^2. Heat flow is the general term measured in kW or W.

1.5.13   Mass Flux and Mass Flow

The term "mass flux" is the quantitative wording used for mass flow per unit area and is usually measured in kg/m^2 s. In this monograph, the term is used to describe the "surface evaporative mass flux".

Mass flow is the general term, which is measured in kg/s.

1.5.14   Liquid Terminology

The following shorthand terminology is used throughout the monograph:

Single component liquids include LIN for liquid nitrogen, LA for liquid argon, LOX for liquid oxygen and LCH4 for liquid methane. Multi-component liquids include LNGs or liquid natural gas mixtures, of mostly methane with ethane and some higher hydrocarbons, and LPGs or liquefied petroleum gas mixtures of mostly propane and butane.

1.5.15 Rollover

The term "rollover" is used generally to describe the uncontrolled, spontaneous, penetrative convective self-mixing of two layers of multi-component cryogenic liquids with initially different densities. The fluid dynamics of this mixing phenomenon is incorrectly described by the word rollover. However, the term is in wide use today and is used in this monograph. Rollover is only one of a number of evaporation instabilities with cryogenic liquids, and is accompanied by a massive increase of evaporation rate over a long period of time (Mode A), measured in hours, or short period of time (Mode B), measured in seconds or minutes.

References

1. Scurlock, R.G.: Storage and Handling of Cryogenic Liquids: The Application of Cryogenic Fluid Dynamics. Kryos, Southampton (2006)
2. Scurlock, R.: Review of rollover and other mixing phenomena in LNG and other cryogenic liquid mixtures. In: Proc. Gastech 2011, Amsterdam (2011)
3. Scurlock, R.: LNG/LPG and cryogenics. Gasworld 39, 44 (2008)
4. Haynes, W.M., Kidnay, A.J., Olien, N.A., Hiza, M.J.: States of thermophysical properties data for pure fluids and mixtures of cryogenic interest. Adv. Cryog. Eng. 29, 919 (1983)
5. Scurlock, R.G.: History and Origins of Cryogenics. Oxford University Press, Oxford (1992)

Chapter 2
Heat Flows in LNG and LPG Cryogenic Storage Systems at 1 Bar

Abstract In a cryogenic storage system, all the A heat inflows through the insulation, are into the stored liquid. The B heat inflows are absorbed by the "cold" in the boil-off vapour.

They include heat transfer by radiation, conduction and convection from the ambient temperature environment of the storage system.

The total A heat inflows are normally balanced by the latent heat of surface evaporation of the boil-off vapour the Boil-off Rate or BOR. There is no boiling as such in a cryogenic storage system.

If some of the heat is stored by heating the liquid, the BOR will be reduced. This stored heat or 'thermal overfill', can with time lead to unstable rises in BOR to release the stored heat.

2.1 Summary of Heat In-flows

1. All cryogenic liquids have boiling points below ambient temperatures, and therefore suffer heat inflows from ambient across large temperature differences, of the order of 150–160 K for LNG and 20–50 K for LPG. All this heat inflow enters the liquid, by a combination of radiation, convection and conduction. This heat is absorbed by primary convection and carries the heat to the surface of the stored liquid. The main heat flow is in a layer at the boundary of the tank wall—but there are additional thermal currents in the bulk liquid if the tank has a low depth but wide diameter. The heat inflow is normally carried out of the liquid via the latent heat of evaporation of the vapour boil-off generated.
2. There are two types of heat inflow. Type 'A' heat in-flows which are absorbed by the evaporation loss or boil-off, and Type 'B' heat in-flows which can be absorbed by the "cold" in the boil-off vapour.
3. Thermally efficient design of a storage system consists of reducing 'A' heat in-flows to a minimum, while ensuring in the design that 'B' heat in-flows are absorbed by the cold boil-off vapour. Furthermore, any conversion of 'A' heat

R.G. Scurlock, *Stratification, Rollover and Handling of LNG, LPG and Other Cryogenic Liquid Mixtures*, SpringerBriefs in Energy, DOI 10.1007/978-3-319-20696-7_2

in-flows into 'B' heat in-flows, by the use of, say, vapour cooled shields, will be beneficial in reducing liquid evaporation.

2.2 Two Types of Storage System

There are in general two types of storage to include in our discussions. They include:

(a) Large flat-bottom, cylindrical, shore-based tanks of 20–100 m diameter, up to 400,000 m^3 capacity, and ship based trapezoidal and spherical tanks of 10–40 m diameter, up to 50,000 m^3 capacity, all using purged gas foam, fibre or powder insulations, and used for large scale storage and transport operations, with LNG and LPG at 1 bar pressure.
(b) Much smaller vacuum insulated, VI, cylindrical pressure containers up to ~4 m diameter and up to 1000 m^3 capacity, at pressures up to 24 bars, all with zero boil-off. These are used for transport and industrial applications with pressurised LNG, as a green replacement fuel for petrol, diesel and bunker fuel.

2.3 Distinction Between 'A' and 'B' Heat In-flows at 1 Bar Pressure

It is important to distinguish between 'A' heat flows into the liquid through the liquid surface and the wetted walls and floor; and the 'B' flows into the vapour through the unwetted walls of the storage vessel (see Fig. 2.1).

The 'A' heat flows are absorbed by evaporating some liquid as boil-off gas and superheating part of the bulk liquid as thermal overfill. On the other hand, most or all of the 'B' heat flows can generally be absorbed by heating the vapour only, i.e.

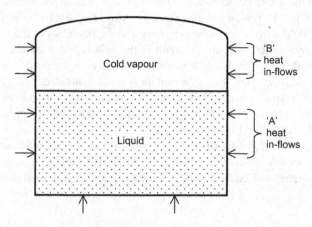

Fig. 2.1 Distinction between A heat inflows absorbed by liquid evaporation, and B heat inflows absorbed by cold vapour

Table 2.1 Comparison of latent heat and sensible heat of cryogens

	T_0 (K)	ρ (kg/m^3)	λ (kJ/kg)	ΔH_s (kJ/kg)	ΔH_s (λ)
Nitrogen	77.31	806.79	198.86	234.5	1.18
Air	78.9	875	202.6	222.3	1.10
Methane	111.69	422.4	510.33	404.0	0.79
Ethylene	169.41	568.0	482.58	182.0	0.38
Ethane	184.55	544.1	488.49	162.0	0.33
Carbon dioxide	194.7*	1560	563	82.0	0.15
Propane	231.1	581	426	107.0	0.25
n-Butane	272.6	601	386	40.5	0.10

by increasing the sensible heat or enthalpy of the evaporated vapour between boiling point and ambient temperature, without adding to the boil-off.

With good design, the 'B' heat flows may not contribute to the evaporation at all, although they may rely on a minimum boil-off vapour mass flow generated by 'A' heat flows.

With the smaller VI pressurizing tanks, there is zero loss of liquid, and no boil-off vapour to carry away the 'A' and 'B' heat flows out of the tank, up to the limiting pressure of the tank. The thermal behaviour of these tanks is discussed later in Chap. 7.

Distinction between the absorption of 'A' and 'B' heat in-flows for different cryogenic liquids at 1 bar can be made by reference to Table 2.1.

For any reasonably well-insulated tank or VI container, the major heat in-flows can be divided into the A flows entering the liquid, and the B flows entering the unwetted areas of the tank or container. The A flows are absorbed by heating and evaporating the liquid; the B flows are absorbed largely by heating the vapour and do not reach the liquid. Both A and B heat flows are a mix of radiation, conduction and convection and include:

- Conduction through the insulation space around the inner containing wall (both A and B flows).
- Convection and radiation within the insulation space (A and B).
- Conduction down the unwetted neck or container walls (A).
- Convection in the vapour above the liquid (A and B).
- Radiation reaching the surface of the liquid from the warmer parts of the container, the roof, the neck, the unwetted walls and through the pipework (A).
- Film flow effects round the container wall above the liquid surface (A).

The ratio of available sensible heat/latent heat varies from 0.79 for methane, down to 0.25 for propane. In general, if the ratio is less than about 0.25, the vapour cooling effect becomes insignificant, and this is so for propane, the higher hydrocarbons and LPGs.

Thermally efficient design of an LNG storage vessel consists, therefore, of reducing the 'A' heat in-flows to a minimum, while ensuring that the 'B' heat inflows are absorbed by the cold vapour and do not enter the liquid. Furthermore, any conversion of 'A' heat in-flows into 'B' heat in-flows will be most beneficial in further reducing the liquid evaporation rate, a trick which can be achieved in several ways described below.

Let us now outline the various heat flows by radiation, conduction and convection respectively, distinguishing between the 'A' heat in-flows and the 'B' heat in-flows. The next chapter, Chap. 3, will go on to describe, in detail, the techniques available for controlling and reducing the heat flows.

2.4   No Boiling

The sum total of the resultant A heat fluxes through the wall and floor insulation are typically less than 15 W/m^2 for LNG and 5.0 W/m^2 for LPG.

It should therefore be emphasised that these heat flux levels are far too small for any nucleate boiling to take place.

For LNG, the level of heat flux through the insulation of 15 W/m^2 is too small by several orders of magnitude, compared with the minimum required to initiate nucleate boiling.

For LPG, the level of heat flux through typical insulation of 5 W/m^2 is too small again by several orders of magnitude, compared with the minimum local heat flux of 10 kW/m^2, required to produce nucleate boiling.

The starting point for discussing the various heat fluxes is therefore the safe assumption that there is no boiling inside an LNG or LPG storage tank or VI container. If you cannot believe this assumption then take a look inside a normally insulated storage vessel using whatever optics or fibre-optics are available. There will be no rising streams of vapour bubbles visible whatsoever; only upward convective motion of suspended solid particles of impurities close to the container wall and, via thermals, at some distance from the wall.

2.5   Overall Convective Circulation in the Liquid

With the total absence of any boiling, all heat entering the liquid, by radiation, conduction or convection, is absorbed by primary natural convection currents which carry the heated (strictly superheated with respect to T_o) liquid to the surface, in an open loop circulation (Fig. 2.2).

At the vertical walls, in vessels with depth/diameter ratios of unity or greater, a relatively high velocity, liquid, boundary-layer flow develops and carries liquid heated by the wall to the surface. At the floor of the vessel, heated liquid in contact with the floor, which has itself been heated from below through the floor insulation, is swept across to join the vertical boundary flow at the walls.

Similarly, in a spherical tank, the primary convection currents are believed to be boundary layer flows at and round the inner curved walls.

At the liquid surface, the superheated wall boundary layer flow turns through 90° (or less or more in the case of spherical containers depending on the position of

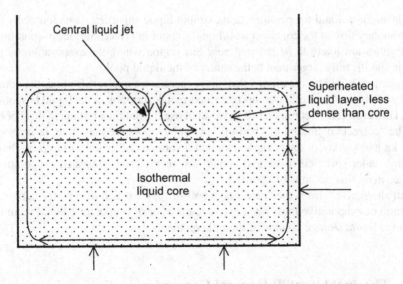

Fig. 2.2 Overall open-loop convective circulation in liquid, producing superheated layer above isothermal core

the liquid/vapour surface) and moves horizontally and radially inwards, just below the surface.

It is during this inward radial flow that evaporation takes place as described in Chap. 4.

At the centre, the liquid motion becomes focussed into a strong downward jet. This central jet carries excess superheat, which has not been released by surface evaporation, into the liquid core as thermal overfill. Secondary convective processes now produce mixing and distribution of the excess superheat either throughout the core, or alternatively in a stratified layer just below the surface. A large depth/diameter ratio will tend to encourage this type of stratification.

For smaller depth/diameter ratios of the order of 0.5 or less (for example, large diameter, cylindrical LNG and LPG tanks) additional convection by a small number of upward thermal plumes in the liquid will carry heat from the floor to the surface. Each thermal is surrounded by a shell of sinking, colder liquid, the two flows constituting a convective cell, rather like a large Rayleigh-Bénard cell, with a diameter approximating to the liquid depth. Similar, much larger, convective cells, around 1000 m diameter, occur widely in the atmosphere, and the central rising thermals are used by glider pilots to gain height as standard practice.

From measurements of the local heat transfer from heated vertical surfaces to cryogens in the natural convection regime, it appears that the laminar boundary layer flow undergoes a transition to turbulence when the modified Grashof number (Gr^*) is of the order of 10^{13}. This agrees with experiments performed with water. For a heat flux of 100 W/m^2, the wall boundary layer can be expected to be turbulent above a liquid height of 0.3 m in LNG, with an increase in heat transfer coefficient for wall/liquid heat transfer [1].

The highest liquid temperature, or maximum liquid superheat, is undoubtedly in the boundary flow at the container wall/liquid-vapour interface. Surface evaporation instabilities are likely to be induced near this region where the evaporation mass flux is significantly larger than at the centre of the liquid pool.

Again, it should be mentioned that, since there is no nucleate boiling, the commonly used term "boil-off" is a short and incorrect description of the liquid evaporation. However, it is in common use so we shall use it from time to time in this book.

The "correct" definition of "boil-off" is total evaporative mass rate, in units of kg/s, kg/h or kg/day, or total evaporative liquid volume rate, in units of m^3/s, m^3/h or m^3/day, under specified (assumed) conditions, or the total integrated evaporative surface mass flux, in units of kg/m^2 s.

An alternative term is the "percentage boil-off/day", which may be the percentage ratio of evaporative mass rate to storage mass, or of evaporative liquid volume rate over liquid storage volume of the full container.

2.6   Thermal Overfill: General Concept

Before discussing the details of heat flows into a cryogenic storage system and the methods for reducing them to acceptable low levels, this chapter starts with building a picture of the overall thermal concept, the absorption of heat flows and their subsequent release by surface evaporation, with no boiling or vapour bubbling [2].

In Chap. 4, we show that the evaporation is driven by part or all of the stored liquid becoming superheated; the greater the surface liquid superheat, the higher the evaporation rate, with no boiling or vapour bubbling.

The overall amount of superheat energy, or "thermal overfill (TO)", is therefore an important parameter for describing the overall thermodynamic state of the stored liquid.

The manner in which TO varies with time is a measure of the stability of the storage system [2].

In formal terms, thermal overfill is the sum of the excess enthalpy $(H - H_0)$ of the stored liquid in relation to the value of H_0 defined for the surface of a homogeneous liquid in thermodynamic equilibrium at T_0, with its saturated vapour at a prescribed pressure P_0. For normal isobaric storage under atmospheric pressure at sea level, the prescribed reference pressure will be close to, but not exactly equal to, 1 bar, and will depend on operational and environmental conditions.

When $(H - H_0)$ is positive for a liquid element, the liquid is described as being "superheated"; when $(H - H_0)$ is negative, the term "sub-cooled" is applicable.

For a large tank, the thermal overfill is defined by:

$$(TO)_{av} = \sum (H - H_0) \tag{2.1}$$

where the summation is over all elements of the stored liquid.

Since $(H - H_o)$ is a function of temperature, density, hydrostatic pressure, composition and thermal history, it can be positive or negative for different elements in the same vessel or tank, in which case,

$$(TO)_{av} = (TO)_+ + (TO)_- \qquad (2.2)$$

where $(TO)_+$ is the sum of the positive contributions, and $(TO)_-$ is the sum of the negative contributions, with a negative sign.

There is frequently poor mixing between different elements of liquid and it is unrealistic to assume that the positive and negative components of $\Sigma(H - H_0)$ cancel each other out.

The important part, $(TO)_+$, is the positive excess energy in the vessel which, if released by the uncontrolled vaporisation of liquid, could lead to an overpressure in the vessel. It should therefore be borne in mind that, when the term thermal overfill is used, the relevant quantity is usually $(TO)_+$ because of the poor mixing.

Before discussing the concept of thermal overfill, let us consider the basic energy equation relating the heat absorbed by evaporation, $m_{(ev)}\lambda$ (where $m_{(ev)}$ is the evaporation mass flow and λ is the latent heat of vaporisation), to the total heat inflow Q through the insulation into the liquid.

Then, under equilibrium conditions, the following energy equation applies:

$$Q - m_{ev}\lambda = 0 \qquad (2.3)$$

Thermal overfill is largely a characteristic describing the liquid "core" which does not take part in the primary convection flow, the latter being driven by heat absorption at the wall and floor, and evaporation at the surface.

Thermal overfill is also time dependent. If the total heat flow into the stored liquid exceeds the heat absorbed by the "boil-off" vapour mass flow rate, $m_{(ev)}$ via the latent heat of vaporisation λ, then the rate of change of thermal overfill with time is:

$$d(TO)_{av} / dt = Q - m_{ev}\lambda \qquad (2.4)$$

For safe storage, the left-hand side of all three equations (2.1), (2.3), and (2.4) should be zero.

If the left-hand side of (2.4) only is zero, then a meta-stable state with constant $(TO)_{av}$, or constant superheat, exists. This is the normal storage situation because, as will be discussed in Chap. 4, a finite, constant superheat or thermal overfill is a necessary condition to drive the equilibrium evaporation of a cryogenic liquid.

However, when the rate of change of thermal overfill $d(TO)_{av}/dt$ is positive over a period of time, then a hazardous storage situation is building up; part of the heat inflow is being stored in the liquid and is not being absorbed by evaporation and removed from the liquid. It should be noted that $d(TO)_{av}/dt$ is positive when the atmospheric pressure (and the reference pressure P_0) is falling.

The magnitude of the thermal overfill can be very considerable . For example, consider LNG in a large storage tank of 100,000 m^3 liquid capacity with bulk super-heats of 0.1, 0.2 and 0.4 K. Then the associated thermal overfills are 14,700, 29,400 and 58,800 MJ respectively; large quantities of energy to dissipate.

Compared with the chemical energy stored in the liquid as heat of combustion, these thermal overfill energies are, however, relatively small; only a few per cent. On the other hand, they are physical energies which may be more easily released than chemical energy via triggering mechanisms. These triggers may not be so eas-ily identified.

With a design boil-off rate of 0.03 %/day of LNG, i.e., 30 m^3 of liquid per day, the equilibrium heat flow into the tank is 75 kW, or 6500 MJ/day.

Using (2.4), the required superheat is only 0.063 K to generate the necessary surface mass flux of 0.074 g/m^2 s to absorb 6,500 MJ/day at an equilibrium thermal overfill of 8800 MJ.

In the example, the LNG has bulk liquid superheats of 0.1, 0.2 and 0.4 K, which together with corresponding evaporative mass fluxes of 0.12, 0.26 and 0.57 g/m^2 s (from (2.4)) are equivalent to boil-off figures of 0.05, 0.1 and 0.23 %/day. These boil-off rates will dissipate thermal overfill at the rates of about 10,500, 22,750 and 49,800 MJ/day, daily figures which are similar in magnitude to the thermal overfills of the bulk superheated liquid.

It therefore follows that the excess thermal overfill from bulk superheats of 0.1, 0.2 and 0.4 K, will be dissipated by excess boil-off, falling from peaks of about 1.6, 3.5 and 7.6 times the normal rate respectively, back to the normal rate over a period of about 1 or 2 days.

2.7 Radiative Heat In-flows

Cryogenic liquids and vapours are generally transparent to infra-red radiation with wavelengths greater than 1 μm, i.e., to nearly all black-body radiation from 300 K and colder sources. In contrast, all plastics, glasses and oxidised surfaces are totally opaque to infra-red radiation at these wavelengths, with reflectivities close to zero (and emissivities close to unity). On the other hand, metals may have very high reflectivities, and emissivities as low as 0.02 (for Al) and ~0.05–0.10 for polished steels.

These low emissivity values for metals can be used advantageously to reduce radiation fluxes as will be described later, but they may rise rapidly and approach unity for oxidised or rough metal surfaces.

Ambient temperature 300 K radiation incident on the liquid surface, an 'A' heat in-flow, will be transmitted through the liquid to, and absorbed at, the inner walls and base of the enclosing vessel. This is perhaps surprising, but with cryogenic liquids having total transparency to infra-red radiation around the 300 K peak, they do not absorb any energy from the radiation.

'B' in-flow radiation incident on the unwetted walls will be totally absorbed if the walls are rough and have a high emissivity of unity. If the walls are smooth and have a low emissivity, then the incident radiation will be reflected down into the liquid as an 'A' heat in-flow.

Radiation in pipework may be a problem since it can be funnelled via reflections at grazing incidence into the liquid as 'A' heat in-flow, with little absorption on the way.

2.8   Conductive Heat In-flows

The main 'A' heat in-flow of conducted heat is through the insulation in the space surrounding the inner liquid vessel. Other sources include conduction down the unwetted wall, through the mechanical supports of the inner vessel, and through the pipework. Conduction through the vapour in contact with the liquid is generally small, but should not be forgotten, particularly when the liquid has a large surface area, and when the vapours have relatively high thermal conductivities.

In very large tanks, insulation below the floor of the tank has to be load-bearing across the whole floor area, and techniques to prevent freezing of the underlying ground have to be used to avoid the serious problem of frost-heave. As a result, the 'A' heat in-flow through the floor insulation tends to be the dominant conduction term.

'B' heat in-flows include conduction down the tank wall from the top, and through the insulation to the unwetted walls; also conduction along pipework entering the liquid from the roof or top plate.

2.9   Convective Heat In-flows

Density stratification in the vapour space over the liquid, would be expected to act so as to prevent convection. In practice, however, there is considerable convective motion of the vapour (see Fig. 2.3).

Heating of the vapour by the wall generates a strong natural convective boundary layer flow up the wall. At the same time, there is a reverse flow in the centre of the vapour column which carries heat as an 'A' in-flow to the surface of the liquid.

The wall boundary layer flow in turn ensures good cooling of the walls by the cold vapour. This vapour cooling effect is a bonus, since it can be used to absorb, as 'B' heat in-flows, part or all of the conducted heat flows down the walls which would otherwise enter the liquid as 'A' heat in-flows.

The heat transfer in the steep vertical temperature gradient in the vapour space is also greatly enhanced, and this aids the efficiency of the vapour cooling effect.

The purpose of most insulation material with a cellular or fibrous structure is to restrict gas convection within it. When faults occur in the insulation such as voids and large cells, then strong convective circulation takes place, with large local heat flows across the cells. In many cases,these insulation faults are revealed as cold spots by ice deposition.

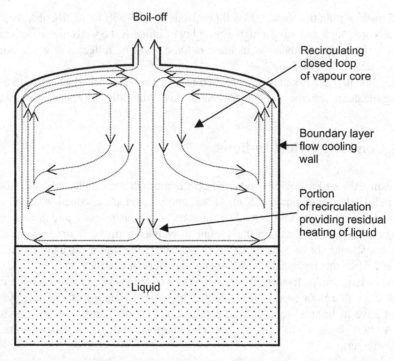

Fig. 2.3 Convective circulation of vapour with recirculating closed loop of vapour core, a portion of which provides residual heating of liquid

2.10   Other Sources of Heat Flow into the Liquid

One additional source of heat flow is that arising from resonant thermo-acoustic oscillations in the vapour columns above the liquid, which can lead to very large 'A' heat flows into the liquid. This problem can be fixed by using a tunable acoustic resonator, tuned to the resonant frequency of the oscillations.

Other sources of heat in-flows include mechanical vibrations, sloshing during transport, eddy current heating and Joule heating of measuring instruments.

References

1. Beduz, C., Rebiai, R., Scurlock, R.G.: Thermal overfill and the surface evaporation of cryogenic liquids under storage. Adv. Cryog. Eng. **29**, 795 (1983)
2. Scurlock, R.G.: Heat flows into a cryogenic liquid storage system. In: Scurlock, R.G. Low Loss Storage and Handling of Cryogenic Liquids. Kryos, Southampton (2006). Chapter 3

Chapter 3
Insulation: The Reduction of 'A' and 'B' Heat In-flows

Abstract All heat inflows are reduced as far as possible by using the correct insulations.

This chapter is concerned with identifying all the A and B heat inflows, and how these can be reduced individually, or collectively, with suitable insulations.

The first type, suitable for large tanks, is the gas-purged insulations such as perlite powder, fibreglass, plastic foams and rock wool. It is important that gas purged insulations totally fill the insulation space between inner and outer containers, with no holes or gaps. Any unfilled space will allow strong convection cells of purge gas to thermally short circuit the insulation. Ingress of water must also be excluded, because the water will freeze to ice, which has a high k value compared with the insulation.

The second type, suitable for smaller tanks, is evacuated insulations. The latest versions are composed of multi-layer reflective insulation MLI with extremely low k values, provided the vacuum is maintained.

3.1 Summary of Insulation Techniques

1. Identify location of all heat inflows and then reduce them to the same magnitude using insulation suitable for scale of operation.
2. In large flat bottom, shore, ship and floating platform tanks, gas purged insulations at 1 bar include perlite and other powders, glass wool blankets or expanded plastic foams, which must completely fill the insulation space. Holes and gaps provide thermal short circuits through the insulation and must be avoided. Smaller bunker fuel tanks (1–1000 m^3 capacity) are vacuum insulated with either powder insulations with medium vacuum required, or multi-layer reflective insulations MLI with high vacuum requirement.
3. k-values for MLI are 10 times lower than evacuated powder and 100 times lower than gas purged insulations at 1 bar.

© The Author(s) 2016

R.G. Scurlock, *Stratification, Rollover and Handling of LNG, LPG and Other Cryogenic Liquid Mixtures*, SpringerBriefs in Energy, DOI 10.1007/978-3-319-20696-7_3

4. Design system to convert 'A' heat in-flows into 'B' in-flows by using vapour cooled baffles, vapour cooling of necks and vapour space walls, and pipework entering liquid.
5. Use metallic radiation shields instead of plastic foam neck plugs in small containers.
6. Note minimum heat inflow to liquid may be determined by reverse convection in the core of vapour columns.
7. For large, shore based, flat-bottom tanks, requiring under-floor heating to prevent frost heave, build in extra thickness of water-proof, load-bearing insulation below the tank-floor.

3.2   Reduction and Control of Heat In-flows

To minimise heat inflow to the stored liquid, it is important to identify all heat sources and then to use available insulation techniques to reduce all their heat inflows to the same order of magnitude, or to zero if practicable. It is important to recognise that variation with scale is different for the various heat sources and therefore the effort and cost of reducing the heat in-flows will also vary with scale.

For ship tanks, it is important to recognise that an additional function of the insulation is required, namely to shield the mild steel, outer hull structure of the ship from being cooled below the steel embrittlement temperature.

The next point to realise is that the available enthalpy for absorbing the various heat-flows is the sum of (a) the latent heat of vaporisation, plus (b) the available cold in the vapour (or the sensible heat increase between boiling point and ambient temperature.

For LNG, the latent heat is about 510 kJ/kg while the sensible heat is about 350 kJ/kg. For minimum boil-off, the A heat in-flows (absorbed by the latent heat) need to be minimised by converting as many as possible into B heat inflows, to be absorbed by the cold vapour and not the liquid.

The first indication of poor insulation, or a fault in the insulation, is the presence of snow or ice patches on the outside of a tank or cryogenic container system. These cold patches indicate the presence of holes in the insulation and should not be tolerated.

The presence of snow on a vapour vent line is also an indication of a poor design which is not making adequate use of the cold vapour in absorbing heat flows.

A number of techniques are available for converting 'A' heat in-flows, entering the liquid, into 'B' in-flows. The most important include:

- Vapour cooled radiation baffles or a suspended deck in the cold vapour space.
- Vapour cooling of the unwetted tank wall to reduce conduction into the liquid.
- Vapour cooled multi-shields to reduce thermal conduction through the wall and floor insulation spaces as A heat inflows into the liquid.

3.3   Radiation

3.3.1   Stefan's Law and Low Emissivity Materials

The major source of infra-red radiation into the liquid is the warm upper parts of the storage tank, and this heat flow is governed by Stefan's Law, which states that the radiation flux between parallel surfaces at absolute temperatures T_1 and T_2, with $T_1 > T_2$, is given by:

$$Q/A = \sigma e_1 e_2 \left(T_1^4 - T_2^4\right) / \left(e_1 + (1 - e_1) e_2\right) \tag{3.1}$$

where e_1 and e_2 are the "total emissivities" of the two surfaces at temperatures T_1 and T_2 respectively, and Stefan's constant, $\sigma = 5.67 \times 10^{-5}$ kW/m^2 K^4.

For 300 K black body radiation, the radiation peak is at 10 μm wavelength in the infra- red, and Q/A is 460 W/m^2 (i.e., for $e_1 = e_2 = 1.0$ and $T_2 \ll T_1$).

For a 75 m diameter tank, the black body radiation inflow from the tank roof into the LNG would therefore be 2032 kW. For a suspended deck temperature of 120 K, the black body radiation inflow would be reduced by a factor of 39.0–52 kW into the LNG.

From the available data on low temperature emissivity materials [1], several generalisations can be made, namely:

- The best reflectors (with the lowest emissivities) are also the best electrical conductors, e.g., silver, e=0.01, gold, e=0.01, copper, e=0.015 and aluminium, e=0.02 (all reflectors at 120 K and exposed to 300 K black body radiation).
- The emissivity *decreases* significantly with decreasing temperature.
- The low emissivity of good reflectors is *increased* by surface contamination, e.g., grease and oxidation.
- Alloying a good metallic reflector *increases* its emissivity, e.g., aluminium alloys have emissivities of 0.06 and higher.
- The emissivity is *increased* by work-hardening the surface layer, e.g., by mechanical polishing to give a shiny finish.
- Visual appearance is not a reliable guide to reflecting power at infra-red wavelengths of 10–100 μm.

We can now begin to see how the 300 K radiation heat flow into LNG can be reduced by a large factor of 500–1000 by using a set of low emissivity baffles, or an equivalent multi-layer aluminium, suspended deck, each cooled by the cold vapour to a temperature of 150 K or less, above the liquid surface.

3.3.2   Vapour-Cooled Radiation Baffles and Suspended Decks

The first systematic studies on the use of vapour-cooled baffles were carried out in the early 1960s at Southampton University, and were reported in 1965 at an IIR conference in Grenoble, France by Lynam et al. [2]. Their paper describes experimental work which showed how ambient temperature radiation may be almost completely and simply absorbed by positioning a set of vapour-cooled horizontal disc baffles above the cryogenic liquid surface.

The disc baffles work in the following way. The downward radiation heat flow from the top of the tank or container is partially absorbed and partially reflected back. The baffles are in turn cooled by increasing the enthalpy of the cold vapour. For each baffle, the radiative heating is balanced by the vapour cooling. In this way, with a series of baffles, the ambient radiation heat flow is almost completely stopped from entering the liquid and contributing to the liquid evaporation.

The radiation, an 'A' heat inflow, has been almost completely converted into a 'B' heat inflow.

We now know that the cooling of the baffles by the rising cold vapour takes place via an efficient heat transfer process, because the natural convective heat transfer for the vapour is greatly enhanced (perhaps by as much as tenfold or more) in the vertical temperature gradient above the liquid.

This simple technique can be applied to all cryogenic systems, including small-scale refrigerated systems, all storage containers for cryogenic liquids, and the large scale storage of LNG, LPG and other hydrocarbon liquids.

In very large LNG tanks, the vapour cooled baffle system is called a suspended deck, which has revolutionised and simplified their design and has, at the same time, significantly reduced construction costs. See Fig. 3.1.

The use of vapour cooled baffles was the subject of several patents back in 1965, but since the technique is so easy to apply, the general use of baffle systems has spread very quickly into all areas of cryogenics, regardless of patents.

There are two simple alternatives to vapour cooled baffles which have been tested, namely plastic foam plugs and floating ball blankets, but both are not so effective.

Fig. 3.1 Vapour cooled suspended deck insulation in upper section of large LNG tank at 1 bar

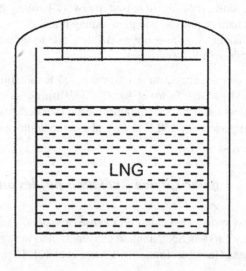

3.3.3 Plastic Foam Plugs

In the original work on vapour cooled baffles published in 1965 [2], expanded poly-styrene or polyurethane foam plugs were demonstrated to be as effective as horizon-tal baffles and this finding led directly to the widespread use of foam plugs. However, subsequent work, published in 1969 [3], demonstrated clearly that foam plugs become unreliable and ineffective insulators after continuous exposure to boil-off gas over a few days.

In other words, when the low thermal conductivity, high molecular weight, foam gas within the plastic foam is replaced by diffusion by high thermal conductivity, low molecular weight methane gas, the foam plug becomes ineffective, behaving as a thermal short circuit in the vapour space.

This thermal short circuiting effect can be easily detected by simply removing the foam plug, when the system will have a lower boil-off rate!!

If a set of horizontal, low emissivity metal baffles replaces the foam plugs, a significantly lower boil-off rate will result. This finding is cost effective although it seems to have been forgotten.

3.3.4 Floating Ball Blankets

Hollow plastic balls, 10–20 mm diameter, either plain or aluminised, are commonly used to reduce evaporation of volatile liquids in storage tanks at ambient tempera-tures. They may also be used to absorb radiation heat flows into cryogenic liquids.

Tests showed that a minimum of 3–5 layers floating on the surface were effective [4]. However, over a long period of time, it is expected that vapour diffusion through the plastic walls will eventually lead to liquid condensing inside the balls. The balls will then sink to the bottom of the tank or, if the layers are stirred up during filling or decanting operations, explode on reaching warmer parts of the tank. These expec-tations, together with the fact that their plastic composition is brittle at low tempera-tures, has led to very little application of this simple idea.

3.4 Conduction Through the Insulation Space

3.4.1 Dewar's Silvered Vacuum Flask or "Dewar"

Conducted heat, whether by gas conduction, solid state conduction or through solid-solid contacts, is reduced by use of the correct type and thickness of insulating materials in the insulation space surrounding the primary liquid container or tank.

Sir James Dewar, at the Royal Institution, London, was among the first to realise in 1877 that he needed better insulation to contain the small quantities of cryogenic liquid, he was producing, for a sufficiently long time to make physical measurements on them. He therefore spent his first 15 years at the Royal Institution on developing cryogenic insulations, before he was later able to liquefy hydrogen in 1898.

He concentrated on evacuated double-walled glass vessels into which he placed a variety of powder, paper and solid materials available to him in the 1880s (at that time, perlite and silica aerogel powders, plastics in foam or any other form, and glass fibre did not exist). Most materials did not work for him, including one consisting of three layers of aluminium foil cigarette packaging [5]. Had he doubled the number of layers, he would undoubtedly have discovered the principle of multi-layer insulation (MLI) some 70 years before Peterson's publication in 1951 — but he missed that discovery [6].

What he did discover by chance, arose from the mercury-vapour vacuum pump he was using to evacuate his experimental double-walled glass vessels. Some mercury droplets had condensed into the vacuum space, on the inside of the outer glass wall. When the inner vessel was filled with liquid air, a shiny mercury film was progressively deposited on the inner glass container because of the finite vapour pressure of mercury at ambient temperature. The liquid air evaporation rate progressively fell to a lower value than with any of the previous insulations he had tested.

Dewar noted this and, in 1892, cleverly replaced the mercury film with a silver coating to achieve the same low evaporation. At last, after 15 years of systematic study, he had mastered the problem of containing liquid air and later, liquid hydrogen, in an evacuated, double-walled, silvered glass vessel or "Dewar".

In fact, he never patented this discovery, and it was his German glass blower, Herr Muller, who developed the idea after he discovered one night that a "Dewar" vessel was also good for keeping liquid hot (the milk for his baby's feeding bottle during the night!). The Thermos Flasche or Thermos flask was born that night, and very soon went into mass production, in the first place as a German product.

Dewar's flask works by cutting out heat transfer, from both A and B heat flows, by conduction and convection with the use of a vacuum in a low thermal conductivity vessel, and reducing radiation by the use of low emissivity reflecting films of silver. It still represents a standard for today, by addressing all the insulation techniques required to minimise boil-off.

3.4.2 Gas Purged Insulations at 1 Bar

There is no doubt that vacuum insulation, with the requirement for an additional outer case strong enough to withstand a collapsing pressure differential of 1 bar, is a complicating nuisance which is avoided for economic reasons on large scale industrial applications of LNG and LPG. It also has to be avoided, because of its weight penalty, for space applications employing LH2 and LOX prior to take-off.

Let us therefore look at insulations with no vacuum, and then consider briefly whether they might be improved. A wide variety of materials are available today, some natural, but most are synthetic (see Table 3.1).

Table 3.1 Effective thermal conductivities of gas-purged and evacuated insulations between 300 and 77 K, in mW/m K

	Pressure (1 bar Nitrogen)	Evacuated to 0.1 Torr
Expanded perlite	26–44	1.0
Silica aerogel	19	1.6
Fibre glass	25	1.7
Foam glass	35–52	–
Balsa wood	49	–
Expanded polystyrene	24–33	–
Polyurethane foam (PUF)	25–33	–
Rock wool	30–43	–

The basic idea of an insulation at 1 bar pressure is to reduce or eliminate heat transfer by gas convection by the creation of sufficiently small gas cells within a matrix of low thermal conductivity powder, solid fibres or foam walls. It follows that (a) holes and gaps provide large cells for convective heat transfer, with k-values many times larger than the insulation and acting as thermal short circuits, and should be avoided, and (b) the insulation material must completely fill the insulation space between outer, ambient temperature tank wall and inner liquid holding tank wall.

At the same time, the insulation is required to be load-bearing, ductile or non-brittle, easy to apply and trouble free in operation, as well as having a thermal contraction to match that of the inner vessel.

To use insulating materials at atmospheric pressure, continuous purging with a water free and non-condensing gas, at a small overpressure, is necessary for two important reasons:

1. To prevent the ingress of water vapour condensing and freezing. The effect of liquid water in the insulation above 273 K is serious enough because water has a high k-value, in the region of 550 mW/mK, which is 10–25 times the k-value of the insulation. However, once the water is frozen to ice below 273 K, the insulation efficiency is even more seriously impaired because ice has a much higher thermal conductivity, with a k-value of 2200 mW/mK at 273 K, which is 40–100 times the k-value of the insulation.

   Furthermore, if the insulation is open to the atmosphere, the freezing of water vapour and formation of ice is progressive, and water vapour will be sucked into the freezing zone to create ice-bridges, which are literally thermal shunts across the insulation. The use of load bearing, rigid foams, such as PUF and glass foam, also lightweight perlite concretes, must rigorously exclude water ingress throughout manufacture, distribution and assembly.

   Care is particularly needed in the design and construction of the foundations of large tanks, to avoid the problems of frost-heave. If the tank is too large to be built on pillars to provide an air gap below the tank, then a platform foundation of insulating bricks of foam glass or perlite concrete needs to be constructed.

The foundation requires an electric heater mat to be laid between two layers of insulating bricks; the whole thickness of the insulation being made 100 % water-tight, particularly if it extends below the local ground-water table.

If water gets in, the k-value for the wet section will increase tenfold; if the water freezes to ice, the k-value will increase 40–100 fold, and the heater-mat may be unable to prevent frost heave, making the installation useless.

2. To prevent any partial condensation of the purge gas through contact with cold surfaces below its boiling temperature at the working overpressure. The build-up of the purge gas condensate within the insulation constitutes a safety hazard to be avoided even with non-flammable insulation, besides introducing an additional heat flow via its latent heat of condensation.

One answer to these problems is to incorporate a water-vapour barrier in the insulation, and use a purge gas at slightly over atmospheric pressure in a gas-tight but not vacuum-tight insulation space. The choice of purge gas is determined by its ready availability and it must obviously not condense at the temperature of the contained liquids.

Dry nitrogen gas purged insulations are commonly used in cold boxes on LNG liquefaction units, storage tanks greater than 3 m diameter, large LNG shore and floating barge tanks, and in sea-going LNG tankers.

Boil-off gas is commonly used as a purge gas in the insulation of land-based storage tanks with suspended deck insulation for LNG, liquid ethylene and liquid petroleum gases (LPG). This works satisfactorily because the boil-off gas is richer in the lower boiling point component, and therefore has a lower condensation temperature than the stored liquid.

For example, an LNG mixture of 80 % methane and 20 % ethane has a boiling temperature of about 114 K, while the boil-off gas will contain more than 95 % methane with a condensing/boiling temperature of say 112 K at the purge-gas overpressure. The margin in temperature is close at 2 K, but acceptable provided the ethane content in the purge gas does not rise above that of the stored liquid mixture (see Fig. 5.4).

The outer cladding of the insulation has to be gas-tight, but since it is under a low pressure differential, it can be built up from thin flat panels to form cold boxes.

In the absence of gas purging in, say, small static systems like transport fuel tanks, ingress of water is a problem which may be reduced by the use of hydrophobic insulation material, as well as by the use of impermeable vapour barriers as part of the insulation. It should be noted that the relatively low k-value of the foam-gas in foam materials, such as expanded polystyrene or PUF, is soon lost when it is replaced through diffusion by purge gas with a higher k-value.

The lower limit of heat flux in a gas-purged insulation is determined by the thermal conductivity of the gas within the cells or matrix of the insulating material. For example, the thermal conductivities or k-values of some typical purge gases at 1 bar are given in Table 3.2.

As can be seen by comparing Tables 3.1 and 3.2, the lowest k-values for nitrogen gas purged insulations are in the region of 25 mW/m K, the quoted value for nitrogen gas at 300 K.

Table 3.2 Thermal conductivities of typical purge gases at different temperatures and 1 bar

	k (mW/mK)	T (K)
Nitrogen	9.6	100
	11.5	120
	25.8	300
Helium	16.9	10
	83.3	120
	156.0	300
Methane	12.8	120
	34.1	300

Table 3.3 Effective thermal conductivities of Multi-Layer Insulations, MLI between 300 and 77 K, in µW/m K

	Layers (cm)	k_{eff} (µW/mK)
0.01 mm aluminium foil + Dexter paper	20	52
0.01 mm aluminised Mylar + Dexter paper	9	200
0.006 mm aluminium foil + nylon net	11	34
NRC2 crinkled aluminised Mylar film	20	28
0.01 mm aluminium foil + carbon-loaded fibreglass paper (Southampton MLI)	30	8–10
0.01 mm aluminised Mylar + carbon-loaded fibreglass paper	30	26–36

For natural gas purged insulations, the lowest k-values are in the region of 34 mW/m K. If the purging fails, and water gets into the insulation and freezes, the k-value could rise to 2200 mW/m K and the BOR rise by up to 60 fold.

In practice, there are problems with powders, and to some extent with fibre blankets, due to thermal contraction and expansion of the inner vessel and pipework. Over a period of time, the insulations settle or move so as to form voids. The presence of such voids are indicated by the appearance of cold spots or frost patches on the outer containing skin. These problems can be avoided with the use of PUF and other plastic foams.

3.4.3 Evacuated Powder Insulations

By reducing the gas pressure to 0.1 Torr, using mechanical rotary vacuum pumps only, the k-value of powder insulations is reduced by a factor of ten or more (see Table 3.3). The gas conduction component has been largely removed. However, a penalty has to be paid since the outer casing must now be vacuum tight and strong enough to withstand a collapsing pressure differential of 1 bar.

With vacuum insulation, the geometry is limited to cylinders, spheres and combinations for mechanical strength, and the size is limited to about 4 m diameter, above which the weight of the outer vacuum-containing vessel becomes prohibitive.

Such vacuum insulated VI tanks with evacuated powder insulation may be used as customer storage tanks, and on LNG fuelled transport such as ferries and road trucks. However, see next section on MLI.

3.4.4   Multi-layer Reflective Insulations (MLI)

Evacuated powders do not address the main residual heat flow through this type of insulation, which is infra-red radiation.

Mixing the powder with low emissivity aluminium or copper powder helps to reduce the radiative heat flow by a factor of 4–5, yielding k-values down to 0.4 mW/m K, as was discovered by Kropschot and his colleagues at NBS, Boulder, USA, in the late 1950s [7].

However, it was found that the wide difference in density between powders like expanded perlite and metal particles of aluminium or copper could lead to separation caused by vibration and packing, with the progressive development of thermal shunts within the powder mixture.

The real breakthrough was made by Peterson in his thesis "The Heat-Tight Vessel" in 1951, when he described replacing the powder with a number of thermally isolated reflecting layers of reflective material, or Multi-Layer Insulation (MLI) [6].

In practice, while there are other contributions to the heat flow, via solid conduction, residual gas conduction and molecular desorption between the reflecting layers, the theoretical limit for MLI is a k-value of 6 µW/mK which therefore represents a target figure to aim for.

A number of different MLIs are now used commercially as indicated in Table 3.3. By using spacer materials between the reflectors, either of fibre glass paper (with no filler) or nylon net, or by having dimples in aluminised plastic film, k_{eff} values down to 30 µW/m K can be realised under test conditions, but somewhat higher values are realised in practice. Nevertheless, these practical values of k_{eff} are at least ten times lower than evacuated powders.

Table 3.3 gives typical quoted values of effective mean thermal conductivities between 300 and 77 K of some MLIs [8].

By paying careful attention to reducing the gas pressure between the reflective sheets, a further threefold reduction in k_{eff} values, down to below 10 µW/mK, can be realised [9]. These lower k_{eff} values may be obtained by the following steps:

(1) Reducing the residual gas pressure in the vacuum space containing the MLI to the minimum by extensive vacuum pumping, and by the strict avoidance of vacuum leaks.
(2) Avoiding molecular desorption within the MLI by high temperature, vacuum baking at 200–300 °C of all the MLI components before assembly or wrapping. Baking at 100 °C will remove adsorbed water, but not adsorbed gases like hydrogen or helium.
(3) Avoiding the use of plastic materials, which contain a great deal of adsorbed molecules and cannot be baked out at sufficiently high temperature to remove them.

(4) Using cryo-pumping materials within the MLI, for example self-pumping spacer material such as activated carbon-loaded fibre-glass paper, to mop up residual molecular desorption after steps (2) and (3) have been carried out. Fibreglass paper, without carbon loading, will also act as a self-pumping spacer after steps (2) and (3), but it has limited adsorption capacity.
(5) Using an additional getter of cryo-pumping materials, within the vacuum space, as a back-up to the high vacuum needed anyway in the depths of the MLI between the reflectors.

3.5   Convective Heat Flows into the Vapour and Liquid

3.5.1   Convective Circulation in the Vapour

In general, density stratification, with hotter fluid, vapour or liquid, above colder fluid, acts so as to oppose vertical natural convection and vertical mixing between layers. In the absence of any heat flow, this convective stability applies to both vapour and liquid.

However, local heating changes the convection picture in a complex manner, with characteristic features which appear to be common to all fluids at temperatures below ambient, including LNG and LPG [10–14]. These features have much in common with convection in the atmosphere, and that is exceedingly complex as we all know.

Basically, in a vapour system, distributed local heating generates a strong upward convective flow at the wall, which is greater than the boil-off mass flow. At the same time, this upward flow induces downward flow in the vapour core.

From continuity of the vapour mass, the following equation applies:

Total downward mass flow = Total upward mass flow − Boil off mass flow

$$(3.2)$$

Alternatively, the vapour convection can be expressed as a recirculating mass flow R(vapour) in addition to the boil-off mass flow.

Studies, of velocity profiles across horizontal planes in vapour columns, using Laser Doppler Velocimetry, have shown that this recirculation mass flow R(vapour) increases with height from a finite value at the liquid surface to a maximum which can greatly exceed the boil-off, and then decreases to zero at the roof of the vessel. Flow visualisation studies have revealed that elements of the core flow are sucked into the wall boundary layer flow at all levels down to the liquid/vapour interface; hence, the variation of R with height and the linear variation of boundary layer thickness with height.

Looking more closely, the downward mass flow of the recirculation is distributed over a large fraction of the diameter of the vapour column, with consequent small velocities. In a narrow neck, the downward flow is localised in a high velocity jet.

3.5.2   Residual Heat Flow from Downward Flowing Vapour

In all cases, the induced recirculating downward mass flow has been discovered to carry heat down towards the liquid/vapour interface. In the case of good insulation practice, this "residual" convective heat flow contributes to the boil-off, and may become the major source of heat entering the liquid. (The term "residual" is used to distinguish this heat flow from all other sources.)

In principle, physically isolating the vapour core from the wall boundary layer flow would be expected to stop the suction of core vapour into the boundary layer. Such a flow isolator would need to create an annular space, adjacent to the wall boundary layer flow, and have zero thermal conductance.

In practice, such a device introduced into the neck or ullage space of a containing vessel or tank has never been observed to reduce the liquid boil-off and is not recommended. Indeed, removal of the flow isolator, and replacement with a set of horizontal radiation baffles has led to significant (i.e. 50 % or more) reduction in BOR for several applications.

Attempts to divert the downward core flow into the upward boundary layer flow above the liquid surface have also never been successful. It should be realised that the vapour recirculation is driven by the heat absorbed from the neck wall by the boundary layer flow, and not by the boil-off mass flow.

3.5.3   Convective Circulation in a Liquid with No Stratification

In the liquid, the heat flow through the container wall is insufficient to produce local boiling. Instead, for depth/diameter ratios of unity or greater, the heat is absorbed by a convective boundary layer flow up the wall to the liquid surface. At the surface, the liquid flow, by this point superheated, turns over through a right angle and moves radially inwards when evaporation occurs as discussed in detail in Chap. 4. At the centre, the flow turns through another right angle forming a strong, downward moving, central jet. This jet is dispersed by mixing with the core (see Fig. 3.2).

Again as in the vapour, the liquid boundary layer flow up the wall induces a reverse, downward flow in the core of the liquid (see Fig. 3.3) such that the following continuity of mass flow equation applies at the liquid surface:

$$\begin{aligned} &\text{Total downward liquid mass flow}(\text{core}) \\ &= \text{Total upward mass flow}(\text{boundary layer flow}) \\ &\quad - \text{Surface evaporation mass flow}(\text{BOR}) \end{aligned} \qquad (3.3)$$

Alternatively, the convective flow can be seen again as a recirculating flow R(liquid) in addition to the heat-inflow generated mass flow. Both heat-inflow generated mass flow and R(liquid) increase with height from the bottom of the liquid, the former reaching a maximum at the surface, while R(liquid) probably reaches a maximum at some point below the liquid surface.

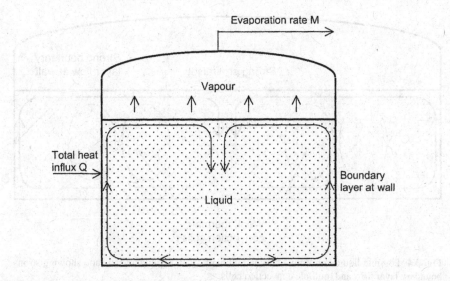

Fig. 3.2 Liquid circulation in storage tank via boundary layer flow at wall. Depth/diameter greater than unity

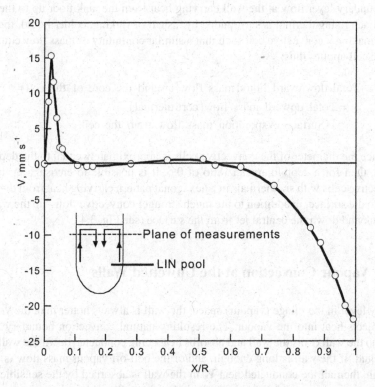

Fig. 3.3 Vertical velocity profile by Laser Doppler Velocimetry in an evaporating LIN pool. X=distance from wall, R=pool radius, depth 13 mm below liquid surface

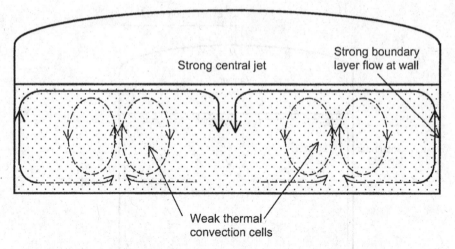

Fig. 3.4 Possible liquid convective flow in tank with small depth/diameter ratio showing strong boundary layer flow and multiple convection cells

For depth/diameter ratios of 0.5 or less, the liquid convective flows induced by the heat inflows are more complex, with additional convective thermals as well as the boundary layer flow at the wall carrying heat from the tank floor up to the surface. Each rising thermal is surrounded by a shell of sinking, colder liquid, the two flows making a convective cell such that a similar continuity of mass flow equation for the cell applies thus:

$$\text{Total downward liquid mass flow(around the core of the cell)}$$
$$= \text{Total upward mass flow(core thermal)}$$
$$- \text{Surface evaporation mass flow from the cell.} \qquad (3.4)$$

Since the diameter of the convective cells is approximately equal to the depth of liquid, then for a depth/diameter ratio of 0.5, it is possible to envisage up to six convective cells with six thermals in a hexagonal pattern carrying heat from the tank floor to the surface, in addition to the much stronger convective flows up the wall of the tank and down the central jet from the surface (see Fig. 3.4).

3.6   Vapour Convection at the Unwetted Walls

At any level in the ullage (vapour) space, the wall is always hotter than the vapour and injects heat into the vapour. The resulting natural convection boundary layer flow up the wall cools the wall and absorbs heat being conducted down the wall into the liquid. If the wall is long enough, and/or the boil-off vapour mass flow is large enough, then all the conducted heat W in the wall is absorbed by the sensible heat of the vapour rising up the wall in the boundary layer flow.

The boundary layer flow is quite different to that normally met at ambient temperatures, where the thickness varies as

$$d \sim z^{0.25} \tag{3.5}$$

and grows very slowly with height z remaining in the region of 0.1–0.2 mm [16].

In the vertical temperature gradient pertaining to the cryogenic vapour column near the liquid surface, the overall convective flow is in the so-called "developed stable region". The boundary layer thickness varies much more rapidly with height as

$$d \sim z \tag{3.6}$$

partly due to the rapid temperature and property (density and viscosity) variation with height and partly through the strong suction of the boundary layer flow drawing elements of vapour from across the whole of the column into the boundary layer.

This suction leads to a progressive increase in boundary layer mass flow and induces the convective recirculation, with accompanying reverse flow in the core of the vapour column. Consequently, the rising boundary layer grows linearly with height and becomes much thicker than ambient temperature boundary layers, with a thickness of the order of 10 mm.

3.7 Other Sources of Heat into the Liquid

3.7.1 Other Sources of Heat into the Liquid

Ambient temperature radiation can funnel down neck tubes and pipelines by internal specular reflection, without significant diminution, directly into liquid baths. Even if the tubes are vapour cooled, the radiation is not absorbed at the walls of the tubes, during internal reflection. To reduce radiation funnelling, the inner surfaces must therefore be rough so as to promote diffuse reflection at the relevant infra-red wavelengths. It is also advisable to use radiation baffles and traps, in all neck tubes and lines entering a cryogenic system.

Holes and gaps in gas purged insulations are mentioned again as providing unwelcome and un-necessary thermal short circuits. Careful filling of the insulation space during construction or manufacture is therefore needed.

3.7.2 Thermo-acoustic Oscillations

It is now widely known that tubes or pipes with large temperature differences along them may contain thermally sustained acoustic oscillations. It is important to avoid these oscillations as they can generate very large "A" heat inflows. These vapour phase oscillations are not adiabatic, or isentropic, because of irreversible heat flows to and from the containing wall, and a net heat current results along the tube towards the cold end. They are named "Taconis oscillations" after Taconis, Leiden University, The Netherlands, first studied these sources of heat into liquid helium in 1949 [15].

Thermo-acoustic oscillations may be accompanied by a low frequency hum, and can be stopped by considering them as an acoustic problem. The solution is a change in geometry at the ambient temperature end and/or the introduction of an acoustic impedance, such as a Helmholz resonator, to damp out the oscillation.

It is interesting to note that the Space Shuttle rocket motor incorporates a cold oscillation damper, called a POGO suppressor, in the LOX pumping line between low pressure and high pressure turbo-pumps [16, 17].

Today, thermo-acoustic oscillations are being used to drive a new class of refrigerator, with no moving parts. This development requires an improvement in understanding wall heat transfer to oscillating vapour columns, which will in turn no doubt help to prevent unwanted oscillations in low-loss cryogenic systems. Thermo-acoustic refrigerators offer a low cost alternative to both pulse-tube and vapour compression, closed cycle refrigerators down to about 100 K.

3.7.3 Mechanical Vibrations

One source of heat flow to mention is that due to induced mechanical vibration. Whether the vibrations are resonant or non-resonant, they are dissipative so that mechanical energy is turned into heat, creating an internal source of heat inflow into the liquid, which may be difficult to identify. The answer is to insulate the cryogenic system from the external sources of vibration, such as vacuum pumps and rotating or reciprocating machinery, by using flexible mountings and couplings.

3.7.4 Eddy Current Heating

Another source of heating can arise from electromagnetic fields causing eddy-current heating in low resistivity materials such as copper, aluminium and superconductors. The answer is to:

(a) divide the materials so that eddy-current loops are minimised,
(b) use high impedance materials, such as plastic composites, where eddy-current heating cannot be induced,
(c) use low impedance electromagnetic screens at ambient temperature surrounding the cryogenic system,
(d) employ combinations of all three.

References

1. Scott, R.B.: Cryogenic Engineering. Van Nostrand, Princeton (1959). 6th reprint (1967)
2. Lynam, P., Proctor, W., Scurlock R.G.: Reduction of the Evaporation Rate of Liquid Helium in Wide Necked Dewars. Bulletin of IIR, Commission 1, Grenoble, Annex 1965-2, p. 351 (1965)
3. Lynam, P., Mustafa, A.M., Proctor, W., Scurlock, R.G.: Reduction of the heat flux into liquid helium in wide necked metal dewars. Cryogenics **9**, 242 (1969)
4. Boardman, J., Lynam, P., Scurlock, R.G.: Reduction of evaporation rate of cryogenic liquids using floating, hollow, polypropylene balls. In: Proc. ICEC3, Cryogenics, vol. 10, p. 133 (1970)
5. Dewar, J.: Proc. R. Inst. **15**, 815 (1898)
6. Peterson, P.: The heat-tight vessel. PhD thesis, University of Lund, Sweden ((1951)
7. Hunter, B.J., Kropschot, R.H., Schrodt, J.E., Fulk, M.M.: Metal additives in evacuated powder insulations. Adv. Cryog. Eng. **5**, 146 (1959)
8. Kropschot, R.H., Schrodt, J.E., Fulk, M.M., Hunter, P.J.: Multilayer insulations. Adv. Cryog. Eng. **5**, 189 (1959)
9. Scurlock, R.G., Saull, B.: Development of multilayer insulations with thermal conductivities below 0.1 μW/cmK. Cryogenics **16**, 303 (1976)
10. Boardman, J., Lynam, P., Scurlock, R.G.: Solid/vapour heat transfer in helium at low temperatures. In: Proc. ICEC4, Eindhoven, p. 310 (1972)
11. Boardman, J., Lynam, P., Scurlock, R.G.: Complex flow in vapour columns over boiling cryogenic liquids. Cryogenics **13**, 520 (1973)
12. Islam, M.S., Scurlock, R.G.: Qualitative details of the complex flow in cryogenic vapour columns. Cryogenics **17**, 655 (1977)
13. Beresford, G.: LDV in cryogenic vapour columns. PhD thesis, Southampton University (1983)
14. Boardman, J.: Heat transfer in vapour columns. PhD thesis, Southampton University (1974)
15. Taconis, K.W., Beenakker, J.J.M., Nier, A.O.C., Aldrich, L.T.: Physica **15**, 733 (1949)
16. Rott, N.: Thermoacoustics. Adv. Appl. Mech. **E20**, 135 (1980)
17. Tward, E., Mason, P.V.: Damping of thermoacoustic oscillators. Adv. Cryog. Eng. **27**, 807 (1982)

Chapter 4
Surface Evaporation of Cryogenic Liquids, Including LNG and LPG

Abstract Surface evaporation of the boil-off gas is the key to understanding how stratification and rollover can take place.

The chapter concentrates on the evaporation mechanism which is controlled by superheated convection currents at the tank wall carrying all the heat inflows up to the liquid surface.

Research studies have shown how the surface evaporation mass flow depends on a thin surface layer with a sensitive thermal impedance which can be strongly influenced by mechanical and convectional disturbances. Rollover and other evaporation instabilities are a consequence.

For LNG mixtures, the surface evaporation impedance of methane is much lower than for ethane, so the vapour has a higher methane content than predicted by standard (T–x) data.

4.1 Summary of Evaporation Processes

1. In a reasonably well-insulated cryogenic liquid storage vessel, all the heat inflow through the insulation from the roof, walls and base of the vessel is carried to the free liquid surface by convection processes.
2. The evaporation mass flux, in terms of mass evaporated per unit area of liquid free surface, varies with bulk liquid superheat ΔT as $\Delta T^{1.33}$ for LNG and LCH4.
3. Conversely, the liquid must be superheated for surface evaporation to take place.
4. Surface evaporation is the only liquid/vapour phase transition taking place, whereby all the heat inflow is absorbed by the latent heat of vaporisation. The enthalpy of the vapour carried out of the vessel balances this heat inflow under equilibrium conditions of storage.
5. Evaporation is controlled by molecular evaporation, mixed thermal conduction/convection, and Rayleigh-Bénard type cellular convection in respectively successive deeper, identifiable, horizontal thin liquid layers, all within a surface sub-layer of about 5 mm below the vapour/liquid interface. The cellular convection is probably driven by surface tension variations in the free surface of the

R.G. Scurlock, *Stratification, Rollover and Handling of LNG,
LPG and Other Cryogenic Liquid Mixtures*, SpringerBriefs in Energy,
DOI 10.1007/978-3-319-20696-7_4

liquid arising from temperature variations created by local differential evaporative mass flows.

6. During normal equilibrium evaporation, these three mechanisms are in a state of fine balance while the evaporation mass flux is determined by the overall temperature difference across these three layers, which is the same as the liquid superheat.

7. The three mechanisms act together as a large, but delicate, impedance between the thermodynamic states of the bulk superheated liquid and the saturated vapour above the surface of the liquid.

8. Any disturbance or agitation of the delicate balance between the three mechanisms will generally lead to a 19 fold increase in evaporation rate for LCH4 and LNG as a vapour explosion over a short time or as a rollover event sustained over a longer period. These increases assume that α the molecular evaporation coefficient remains at 10^{-3} for both LCH4 and LNG. If α is ten times larger, the increase in evaporation rates during a vapour explosion, or rollover, will be ten times larger, i.e., 190 fold.

9. The three mechanisms appear to be self-repairing, within seconds after the disturbance, or agitation of the surface sub-layer, ceases.

4.2   Energy Balance Between Heat In-leaks and Boil-off Rates

All liquids with normal boiling points below ambient temperature suffer from heat in-leaks by conduction, convection and radiation through the surrounding insulations. In practice, the insulations are very efficient in keeping the boil-off rate, BOR, down to a fraction of 1 % per day to reduce loss of liquid during storage, transport and handling. Under normal storage conditions, an energy balance should exist whereby the total sum of heat in-leaks is absorbed by the latent heat of evaporation of the boil-off, i.e.:

$$\text{Heat in-leak, kW} = \text{Normal BOR, kg / s} \times \text{Latent heat of evaporation, kJ / kg}$$

If for any reason, the BOR falls below the total heat in-leak, the thermal energy flow into and out of the liquid is not balanced and the surplus energy flow is stored in the liquid as superheat or thermal overfill. This surplus energy remains in the liquid and builds up with time. It can only be released by evaporation of some of the now superheated liquid via one of the unstable evaporation phenomena described below. These uncontrolled evaporation phenomena constitute a hazard which increases with time, and with the scale of the storage container and with the volume of liquid being stored.

The observed puffs of vapour from a laboratory sized container of LNG represent a major uncontrolled vapour release in a 100 m diameter tank.

4.3 The Modes of Heat Transfer in "Boil-off" Mechanisms

This chapter now describes how there are three modes of heat transfer within the fluid mechanisms which lead to evaporation of the liquid. These are:

1. Nucleate boiling from wall to bulk liquid, at high heat fluxes.
2. Convective heat transfer from wall to bulk liquid with no boiling, at the low heat fluxes characteristic of normal LNG storage.
3. Conductive and micro-convective heat transfer in the liquid/vapour interfacial region leading to surface evaporation as the only normal boil-off mechanism in LNG.

The chapter then continues with a discussion on how, when surface evaporation is dominant, a superheated state of the liquid is a necessary occurrence.

At the same time as steady-state surface evaporation takes place, unstable evaporation phenomena are common occurrences which need to be understood, controlled and accepted. This chapter will conclude with outlines of **four** of these phenomena, namely:

- normal, irregular surface evaporation
- single and multiple vapour explosions via transient high surface evaporation rates, for example the crackling and spitting produced by sloshing of LNG in tanks, and spills of LNG on water.
- Rollover, (so called), with high surface evaporation rates, as a possible result of spontaneous mixing following stratification into two or more layers—a phenomenon that may be met with LNGs, LPGs and other cryogenic liquid mixtures.
- explosive boiling throughout the bulk liquid as quasi-homogeneous nucleate (QHN) boiling.

4.4 Nucleate Boiling from Wall into Bulk Liquid

4.4.1 Heterogeneous Nucleate Pool-Boiling

The mechanism of heterogeneous nucleate pool-boiling on a submerged heated wall is well documented [1]. Because vapour bubbles have an increased internal vapour pressure (proportional to the ratio of surface tension/bubble diameter), they have an increased saturation temperature which must be exceeded for the bubble to grow. The increase in wall temperature needed to create vapour bubbles in the first place can be reduced by nucleation centres in, or on, the surface of the wall.

Heterogeneous nucleate boiling on plain heated surfaces can be significantly enhanced today by treating the surfaces so as to create a wide variety and a high density of nucleation sites, for example, by the application of porous coatings [2].

For LCH4, heterogeneous nucleate boiling on plain surfaces occurs between heat fluxes of a minimum of about 10 kW/m², rising to a maximum or critical heat flux of about 500 kW/m², with wall superheats from about 0.5 K to a maximum of about 20.0 K, respectively, the actual values depending on the particular surface and its immediately previous, thermal history.

For extended vertical surfaces, like those in evaporators or the walls of large tanks, the integrated heat transfer into the liquid over a large vertical distance depends on several factors. These include whether the heat transfer takes place at constant wall temperature (the evaporator condition, which is very difficult to simulate experimentally) or constant wall heat-flux (the usual laboratory experimental simulation), and whether the bulk liquid temperature is constant with depth, or not.

For LNG, T_s rises by 1 K at a liquid depth of 20 m. Nucleate boiling will therefore tend to be suppressed at lower levels in the liquid, particularly under the constant wall temperature condition in a liquid evaporator, rendering the lower part ineffective for heat transfer purposes.

This suppression of nucleate boiling with liquid depth is not observed with constant wall heat flux, experimental boiling heat transfer rigs. A sophisticated computer controlled, variable heat flux with height, constant wall temperature, experimental rig will, however, demonstrate quite dramatically the boiling heat transfer ineffectiveness of the lower part of the rig in an isothermal pool of cryogenic liquid [3, 4].

For heterogeneous nucleate boiling in a liquid at its saturation temperature T_0, streams of vapour bubbles rise to the surface and break through to become the boil-off vapour mass flow.

If the liquid is subcooled with respect to T_0, the rising vapour bubbles may collapse due to recondensation of their vapour content, so that there is no boil-off vapour. Instead, all the latent heat of condensation of the bubbles goes into heating the bulk liquid.

4.4.2   Homogeneous Nucleate Boiling

In the absence of nucleation sites, a liquid can be heated by heat fluxes normally associated with heterogeneous nucleate boiling to a temperature well in excess of T_0 without boiling. With continued heating of the liquid, the excess superheat can become much larger than that for heterogeneous nucleate boiling and large enough for homogeneous nucleate boiling to take place. Vapour bubbles then form spontaneously throughout the liquid and grow very rapidly, almost explosively. Large volumes of vapour mixed with liquid are produced, and a mixture of vapour and liquid may be carried out through the vents and the safety valves. The associated pressure rise due to the vapour generation may damage the liquid enclosure vessel unless the vents have been sized appropriately. In some cases, two-phase flow may be considered, when sizing relief systems.

For LNG, the homogeneous nucleate boiling temperature corresponds to an excess unstable superheat of possibly 40 K for homogeneous nucleate boiling to take place in the violent fashion described above. This violent boiling and crackling, called Rapid Phase Transfer or RPT, is met and/or heard when LNG is spilled on to water.

4.4.3   Quasi-homogeneous Nucleate Boiling, QHN Boiling

In many situations, similar violent boiling can take place with much smaller unstable superheated states than those required for homogeneous nucleate boiling, e.g. for LNG and LPG with superheats of only 1 or 2 K. These events are termed quasi-homogeneous nucleate (QHN) boiling events, and take place in the absence of solid surfaces with nucleation sites [5].

Bubble nucleation during QHN boiling is believed to take place on either

(a) particulate suspended in the liquid, such as CO_2 snow, or H_2O ice,
(b) vapour trails created by the passage of cosmic radiation particles through the liquid, or
(c) focussed sound waves, or pressure oscillations from pulsed acoustic sources, such as pumps, valves, and instrumentation.

In all cases, when QHN boiling takes place, there is a large volume of vapour generated throughout the liquid; and a rapid rise in the liquid/vapour interface, with the possibility of liquid being carried out of the vapour vent lines.

In the case of small containers like bunker fuel tanks, this flow of two-phase mixture of liquid and vapour out of the vent is similar to that of geysering events and may empty the container of liquid.

If the rate of vapour generation exceeds the choked flow discharge capacity of the vent lines and safety valves together, the pressure will rise rapidly until the enclosure fails mechanically.

There is a major problem in detecting the build-up of unstable superheated states in the liquid, or assessing in advance the magnitude of a QHN boiling event, if the liquid temperature is not adequately monitored. This is because the vapour pressure over the liquid is no guide to the degree of superheating of the liquid (see Fig. 1.1). A fall in evaporation rate below the "normal" storage value could be a possible warning guide. Sometimes this is the ONLY warning of a build-up towards a QHN boiling event.

As an example of a QHN boiling event, a pressurising LIN cryostat experiment, at Southampton University for Masters students, accidentally demonstrated this nasty problem very clearly on several occasions. In this experiment, about 2 L of LIN were heated in a 4 L, closed volume, vacuum insulated vessel, via a 30 W wall heater consisting of wire wound around the outside of the pressure vessel, so as to measure the (P, T_0) saturation curve, up to a pressure of 5 bar. The heater wire was not in direct contact with the LIN and presumably its heating effect did not generate any heterogeneous nucleate boiling. Normally, both vapour pressure and liquid temperature increased smoothly with time yielding the standard saturation curve.

However, on several occasions, energising the wall heater caused the LIN to superheat into the unstable superheated state, without any increase in pressure. If the students did not appreciate quickly the significance of their observations of rising temperature but with no corresponding rise in pressure, then after some 30 min, the cryostat would be in a dangerous condition.

Recovery was achieved by turning off the wall heater, evacuating the laboratory, opening the vent valve, and running. The action of opening the vent valve, however slowly, generated a QHN boiling event; the vapour pressure gauge was observed to go off scale above 30 bars, in agreement with the following simple calculation.

A heat input of 30 W into 2 L of LIN for 30 min will increase the enthalpy of the system by 54 kJ. This corresponds to a liquid superheat of about 13 K in 2 L to a liquid temperature of 90 K. This heat is absorbed by the flash evaporation of some 200 L of vapour at 1 bar. If the evaporation is by QHN boiling in a very small time, of, say, less than a second, then it will result in a pressure of 30–100 bars being generated in the closed volume of 2 L before the safety valves can open!

This unpleasant QHN result was repeated using LIN on several occasions by my students. It was not repeated with LNG or LCH4 for safety reasons.

If you would like to repeat this experiment with any cryogenic liquid, then take it out of doors into a remote bunker. Stand well clear because if the container explodes, parts of it may take off like a rocket.

4.5   Convective Heat Transfer Without Evaporation at the Point of Heat Influx

In most, if not all, storage situations, the heat flow through the insulation and tank walls into the liquid is a much more gentle process with a heat flux of typically less than 100 W/m² for LNG. This level of heat flux is some two orders of magnitude less than the minimum of 10,000 W/m² required for heterogeneous nucleate boiling. It can only be released from the liquid by surface evaporation, with no boiling at all.

To get to the surface, the heat in-flow is first absorbed by a process of natural convective heat transfer creating an upward flow of less dense superheated liquid; there is no boiling and also no evaporation at the point where the heat is absorbed. At a vertical wall, the flow of superheated liquid assumes the form of a boundary layer immediately adjacent to the wall, in a layer about 1–5 mm thick. Heat transfer from the heated wall to the liquid by such a boundary layer flow is very effective and is well documented in many texts on heat transfer [6, 7].

For LNG and LPG, the boundary layer flow at the wall absorbs all the heat flow entering the liquid. Furthermore, for a container or tank which has a liquid depth/diameter ratio greater than about 0.5, the heat flow through the base is absorbed convectively by a boundary layer flow across the base which is continuous with the vertical wall boundary layer flow via a boundary layer suction process. The same is believed to be true for spherical containers and tanks, i.e. The boundary layer flow follows the vertical, inner curve of the tank wall.

As mentioned in Chap. 3, for depth/diameter ratios less than 0.5, such as in large cylindrical LPG and LNG tanks, the boundary layer flow across the base may be broken by the creation of vertical "thermals" spaced horizontally at intervals approximating to the liquid depth according to Rayleigh's instability criteria for natural convection. In all cases, the heat in-flow is carried by boundary layer flows, and thermals, to the liquid surface.

4.6   Surface Evaporation, the Only Way of Dissipating Superheat and Thermal Overfill

After the superheated boundary layer flows and thermals reach the surface, some, or all, of the excess heat is absorbed by the process of surface evaporation. In fact, the wall boundary layer flow turns over into a horizontal flow and moves radially inwards just below the liquid/vapour interface. During this inward radial motion, surface evaporation takes place as described in Sects. 4.6.1–4.6.5 below (see also Fig. 4.1).

When the inward radial motion reaches the centre (of a cylindrical vessel or tank), it turns over to become a strong downward jet carrying the excess heat, not released by surface evaporation, into the core of the liquid where secondary convection produces mixing and superheating.

The highest fluid temperature is undoubtedly at the wall–liquid–vapour interface, where the evaporative mass flux will be larger than at the centre of the liquid pool. The mechanism of surface evaporation during inward radial flow is extraordinarily complicated and sensitive, while the rate of evaporating mass flow through the surface is controlled by several local factors. Let us develop the picture of this mechanism as it was discovered at Southampton.

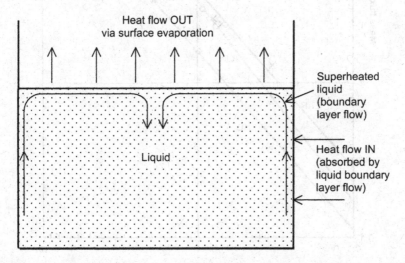

Fig. 4.1 Surface evaporation, whereby heat flow into liquid at the wall is rejected remotely via vapour from the surface

4.6.1   Surface Evaporation Mass Flux and Bulk Superheat

To gain an understanding of the relationship between evaporation mass flux and bulk superheat, microthermometer studies were first made on LIN, LCH4 and LNG [8–11].

The boil-off vessel was an 80 mm inner diameter, double walled, vacuum insulated dewar surrounded by a second liquid bath 120 mm inner diameter. The boil-off from the inner vessel could be varied via a uniform heat-flux electrical heater mounted in the vacuum space around the inner wall. The micro-thermometers consisted of 25 μm diameter copper/constantan thermocouple junctions mounted horizontally in differential or absolute configurations (see Fig. 4.2).

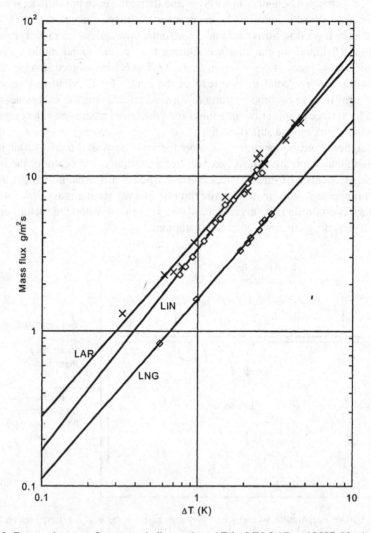

Fig. 4.2 Evaporation mass flux versus bulk superheat ΔT for LIN, LAR and LNG (North Sea)

For the differential configuration, the two junctions were separated vertically by a distance of 100 mm, so that temperatures in the region of the surface could be measured relative to that of the bulk liquid. For the absolute configuration, the reference junction was the ice-point while a single calibration point was made against a Platinum thermometer in rapidly boiling LIN. The pool depth was kept between 200 and 250 mm.

The results are summarised in Fig. 4.2 as a log-log plot of average mass-flux through the surface in g/m² s against bulk superheat ΔT ($T_b - T_0$). With the uniform heater used, it proved possible to generate superheats of 3.0 K in LIN and LNG, before any nucleate boiling occurred.

$$\text{For LIN} \quad m = 3.2 \Delta T^{1.80} \, g / m^2 s \tag{4.1}$$

$$\text{For LNG(North Sea)} \quad m = 1.60 \Delta T^{1.33} \, g / m^2 s \tag{4.2}$$

4.6.2 Impedances to Surface Evaporation: The Three Regions in the Surface Sub-layer

The impedance mechanisms are revealed by the vertical temperature profiles and their variation with time within a few mm of the liquid-vapour interface, which we shall call the surface sub-layer.

To produce these time dependent profiles, the liquid surface was allowed to fall and pass the rigidly fixed micro-thermometer system as the liquid evaporated. An example of the vertical profiles in LIN is shown in Fig. 4.3 for a mass flux of 12 g/ m² s, equivalent to a heat flux of about 2.4 kW/m² through the surface. Similar vertical profiles were observed in surface sub-layers of all the liquids, namely LIN, LCH4 and LNG (North Sea).

As can be seen in Fig. 4.3, the surface is cooler than the bulk liquid while the actual vertical temperature profile has a complex structure in the sub-layer down to a depth of about 5 mm, below which the temperature is uniform at T_b.

These observations are not unique to cryogenic liquids. Oceanographers have a problem with determining sea-water bulk temperatures from infra-red scanning radiometers on satellites. The satellite sensors measure the sea-surface skin temperature, characteristic of the top layer less than 100 μm thick, which can be up to 1 K cooler than the bulk on a calm night in the absence of solar heating [12].

Returning to Fig. 4.3, the profile within the sub-layer is dominated by the superposition of many temperature pulses, or spikes, which become observable with a fast response recorder. Expanding the time scale, these spikes can be seen to show an irregular time dependence and include short cold temperature pulses (but never colder than T_0) and longer hot temperature pulses (but never hotter than T_b). However, the time independent part of the profiles can be separated into three

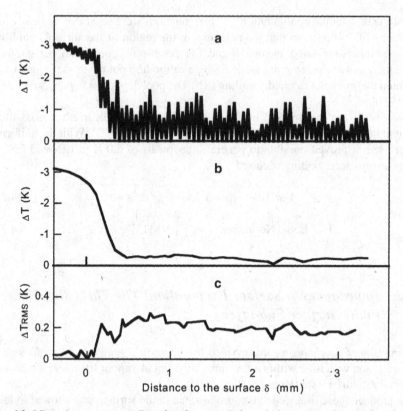

Fig. 4.3 Micro thermometer studies of surface morphology during evaporation. (**a**) Local temperature variation $\Delta T = T - Tb$ with depth δ below liquid surface. (**b**) Smoothed variation with δ. (**c**) RMS variation of fluctuations with δ

regions (working down from the surface) with different temperature gradients, as shown in Fig. 4.4, namely:

(1) A molecular evaporation region at the surface which is probably no more than 1–2 μm in thickness, but appearing to extend to 50–100 μm in practice as the capillary film remains attached to the thermocouple junction by surface tension forces when the liquid surface falls below it.

(2) A thermal conduction region enhanced by some convection, about 400 μm thick, with an extraordinarily high temperature gradient.

(3) An intermittent convection region, about 5000 μm or 5 mm thick, with a small temperature gradient, which contains the bulk of the observed thermal spikes. The thermal spikes also extend into region 2, but rapidly reduce in intensity as the surface is approached.

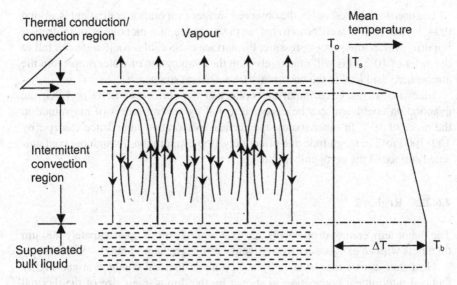

Fig. 4.4 Morphology and temperature profile across the three regions of surface sub-layer on LIN

4.6.2.1 Region 1

At the vapour/liquid interface of 1–2 μm thickness, molecular evaporation takes place according to the relation [13]:

$$M = 90\alpha \left(T_s - T_0\right) kg\,/\,m^2\,s \tag{4.3}$$

This is equivalent to a heat flux Q_{ev} given by:

$$Q_{ev} = m\lambda = 90\lambda\alpha\left(T_s - T_0\right) kW\,/\,m^2 \tag{4.4}$$

where T_s is the temperature of the vapour/liquid interface, T_0 is the saturation temperature, λ is the latent heat of evaporation and α is the molecular evaporation coefficient. The reciprocal of α can be regarded as an evaporation impedance factor, the magnitude of which is very sensitive to the concentration of impurities in the surface.

Figure 4.4 shows the temperature profile across the three layers between the vapour side of the interface at T_0 and the bulk liquid at T_b for LIN at $m = 12\,g/m^2\,s$ where the bulk superheat ($\Delta T = T_b - T_0$) was 3.0 K.

The estimated value of $(T_s - T_0)$ is 0.15 ± 0.10 K, close to the lower limit of measurement. Inserting values into (2.5), it can be seen that this measurement leads to an estimated minimum value of the evaporation coefficient α of about 10^{-3} for liquid nitrogen.

For uncontaminated water, the observed surface evaporation coefficient is around 0.04. However, it is well known that, in the presence of a monolayer of molecular impurity with a low vapour pressure, the surface evaporation coefficient can fall to the order of 10^{-6}. This will effectively stop the evaporation of water droplets in the atmosphere, and lead to the occurrence of persistent smogs.

Likewise, if insoluble impurities collect at the surface of LNG or LPG, the evaporation coefficient can be expected to fall by several orders of magnitude to the order of 10^{-6}. In other words, the surface evaporation impedance can rise by 100–1000 fold in magnitude and effectively stop evaporation through the contaminated surface of the cryogenic liquid.

4.6.2.2 Region 2

The major temperature drop occurs across a liquid layer of approximately 400 µm thickness with an extraordinarily high temperature gradient of 5000–7500 K/m.

Heat flow across this layer is by a mixture of thermal conduction and highly damped intermittent convection as shown by the diminishing size of the thermal spikes on approaching the surface.

The effective heat flux for any liquid is given by:

$$Q_{ev} = m\lambda = k_{eff}\left(dT/dz\right) = k_{eff}\left(T_c - T_s\right)/\delta(2) \tag{4.5}$$

where $\delta(2)$ is the thickness of region 2. For the experimental example with LIN (see also Table 4.1), the value of k_{eff} is 370 mW/mK compared with the static conductivity of 133 mW/mK. Thus we see that the conductance across the region 2 is enhanced by the intermittent convection by a factor of about 2.8.

Table 4.1 shows how this degree of enhanced conductance across region 2 in the evaporative sub-layer applies to the other cryogenic liquids studied, varying from 2.8 for LIN, down to 1.5 for CH4 and LNG.

The heat flow across region 2 for the LIN example is then:

$$m\lambda = 0.8\left(T_c - T_s\right) kW/m^2 \tag{4.6}$$

where T_c is the temperature at the bottom of this layer of mixed conduction and intermittent convection.

Table 4.1 Experimental data on effective thermal conductivity in region 2 of surface sub-layer during surface evaporation

	Surface mass flux (g/m² s)	ΔT (K)	Temperature gradient (×10³ K/m)	k_{eff} (mW/mK)	k_{actual} (mW/mK)	k_{eff}/k_{actual}
LIN	13.9	3.2	7.5	370	133	2.8
LCH4	3.0	2.5	5.0	305	189	1.6
LOX	9.8	2.4	6.7	305	152	2.0
LA	14.6	3.2	10.0	236	128	1.8
LNG	2.7	2.1	5.0	275	189	1.5

4.6.2.3   Region 3

In region 3, the depth interval 0.2–5.0 mm, the temperature difference across this relatively thick layer is only 0.25 K corresponding to a mean temperature gradient of about 50 K/m, much smaller than in the region 2.

The major feature is the temperature spikes which have a time constant of about 0.3 s and a mean amplitude of approximately +0.5 K and –0.5 K respectively within the temperature interval T_b to T_s. For liquid depths $\delta < 0.4$ mm (i.e. <400 µm) within region 2, the temperature spikes decrease in amplitude to zero as the surface is approached at $\delta = 0$. For $\delta > 5$ mm, at depths below the surface sub-layer, the temperature spikes appear to cease.

It is deduced from these profiles, and subsequent video—Schlieren observations (see below in Sect. 4.6.4), that these temperature spikes arise from convection cells carrying heat across the 5 mm thickness of region 3 under the small temperature gradient observed. The cells appear to consist of narrow falling plumes of cold liquid and wider rising plumes of hot liquid co-ordinated into convecting sheets or streamers moving parallel to the surface past the stationary microthermometers. The streamers have a vertical dimension of 4–5 mm, i.e. the thickness of region 3, and are driven primarily by the cooling and sinking, through increase in density, of "spent" elements of liquid, from the mixed conduction/convection region 2, in the temperature gradient of about 50 K/m.

At the same time, through mass continuity, the streamers incorporate elements of superheated liquid from immediately below region 3 which rise and are transferred to region 2 where they give up their superheat.

This lower thick region, region 3, which we call the intermittent convection layer, therefore contains the mechanism whereby superheated liquid from the inward radial flow of the bulk convection loop (driven by the wall boundary layer flow) is carried into the surface layer by a collective system of rising and falling plumes or convective streamers. This mechanism is not unique to cryogenic liquids and was first identified by Rayleigh from his studies on the evaporation of water. He was seeking to understand how superheated bulk liquid can flow convectively upwards to the colder surface, against the negative temperature gradients and positive density gradients.

4.6.2.4   Examples of Total Surface Evaporation Impedance

For the LIN example, the effective linear heat transfer equation for this intermittent convection layer is given by:

$$Q_{ev} = k_{eff}\left(T_b - T_c\right)/\delta\,(3) \tag{4.7}$$

where $k_{eff} = 20$ W/m K, $\delta(3)$ is the thickness of region 3, equal to 5 mm, and T_b is the bulk temperature of the superheated liquid. Hence, for the example,

$$Q_{ev} = m\lambda = 9.6\left(T_b - T_c\right) \text{ kW}/\text{m}^2 \tag{4.8}$$

it can be seen that the effective thermal conductance of this layer due to the convection process, is about 360 times the thermal conductivity of the static liquid. Combining (4.6), (4.8), and (4.10), we obtain the overall heat flux equation for the surface evaporation of liquid nitrogen in the example, where $\lambda = 199.0$ kJ/kg and the evaporation mass flux, m is in kg/m²:

$$(Tb-To) = (Ts-To) + (Tc-Ts) + (Tb-Tc)$$
$$= m[1/90\alpha + 250 + 0.02] \qquad (4.9)$$
$$\quad\;\; (1) \qquad\quad (2) \qquad\quad (3)$$

For the liquid methane example in Table 4.1, the surface evaporation impedance equation, with $\lambda = 512$ kJ/kg and effective thermal conductivity (in region 2) $k_{eff} = 0.30$ W/mK compared with the static value of 0.19 W/mK, becomes:

$$(Tb-To) = m[1/35\alpha + 545 + 0.1] \qquad (4.10)$$
$$\qquad\quad\;\; (1) \qquad (2) \quad (3)$$

The impedance terms represent respectively the contributions from:

(1) Molecular evaporation including impurity effects in region 1.
(2) The mixed thermal conduction/convection layer, region 2.
(3) The intermittent convection layer, region 3.

4.6.3   Summary of Conclusions on the Separate Impedance Terms During Surface Evaporation

(a) Even if the evaporation coefficient α is as small as 10^{-3}, the impedance term (1) is still over 20 times smaller than term (2). In other words, the molecular evaporation process is not the rate limiting mechanism by at least an order of magnitude during normal surface evaporation.
(b) The steep temperature gradients associated with the mixed thermal conduction/convection region 2 are approximately the same (to within a factor of 2) for the various liquids as shown in Table 4.1. The observed thickness of the mixed conduction/convection layer is about 0.4 mm, or 400 µm, for every liquid.
(c) The intermittent convection in region 3 is a common feature of all the cryogenic liquids studied, including LIN, LA, LOX, LCH4 and LNG (North Sea natural gas).
(d) The important conclusion is that the major impedance to evaporation is provided by region 2, the mixed conduction/convection region.
(e) Convection via intermittent plumes in region 3 is much more efficient, with an effective thermal conductance over 1000 times greater than that in region 2, for LCH4.
(f) It needs to be firmly appreciated that this equilibrium process, whereby bulk superheated liquid is separated from the surface by a thin sub-layer in which the heat flux to the surface is almost totally controlled by a mixed conduction/convection layer only 400 µm in thickness, is sensitive to being dramatically

disturbed in two totally different ways, firstly, by agitation of the bulk liquid and, secondly, by impurities in the surface.

(g) Agitation of the bulk liquid can lead to liquid motion in the surface sub-layer circumventing or bypassing the mixed conduction/convection region. In this case the rate limiting impedance falls to that of the molecular evaporation, and, assuming the molecular evaporation coefficient α is 10^{-3}, the overall evaporation impedance **reduces** from ~625 to ~28 resulting in an evaporation mass flux **increase** of 24 fold, for LCH4. For the La Spezia rollover, the increase in evaporation mass flux was 250 fold, suggesting that the value for α was 10^{-2} for that particular LNG.

(h) This is believed to be what happens to the boil-off rate during a vapour explosion over a short period of the order of one second, or during a rollover if the agitation of the surface sub-layer due to penetrative convective mixing is maintained over a long period of 10–1000 s, or more.

(i) Impurities collecting on the surface, either by condensation from the ullage vapour, or from solute remaining behind during evaporation of the solvent liquid, can cause the evaporation coefficient α to fall to the order of 10^{-6}. In which case, the BOR limiting impedance is determined by the reduction in α and **rises** by 40 fold in the case of LCH4, causing the evaporation to almost stop. Possible candidate impurities include water and carbon dioxide.

(j) The surface sub-layer is strongly self-repairing. When the agitation ceases or is removed, the equilibrium 3-layer structure appears to be quickly re-established, within a few seconds.

4.6.4 Schlieren Studies of the Surface Sub-layers

A cryogenic Schlieren system shown in Fig. 4.5 was used in conjunction with a TV camera and video recorder to observe the convective motion within the surface sub-layer directly.

A parallel beam of white light from a point source is directed onto a stainless steel mirror submerged 10–30 mm deep in the cryogenic liquid, i.e. sufficiently deep to be clear of both the surface evaporation sub-layer and the radial inflow across the surface of superheated liquid from the wall heated boundary flow. The reflected beam of white light passes out and is focussed on to a viewing screen or camera, after passing a knife-edge to remove the undeviated beam.

Bearing in mind that the vertical-axis Schlieren optics detect horizontal gradients in refractive index, which relate directly to changes in density produced by horizontal gradients in temperature, local convection patterns are revealed in extraordinary detail.

Figure 4.6 illustrates the instantaneous pattern of radial convection lines or streamers which are revealed. These lines are easily observed on the video recordings and can be seen to be in constant motion, sweeping over the whole surface and moving radially inwards just below the liquid surface, from the peripheral wall

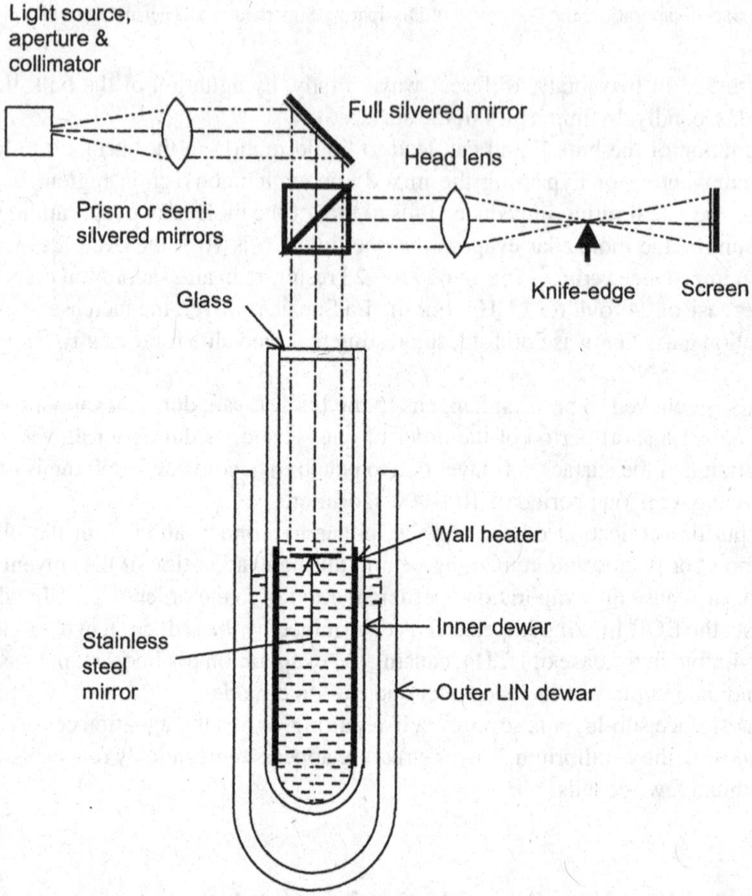

Light source,
aperture &
collimator

Full silvered mirror

Head lens

Prism or semi-
silvered mirrors

Glass

Knife-edge   Screen

Wall heater

Stainless
steel
mirror

Inner dewar

Outer LIN dewar

Fig. 4.5 Experimental set-up for Schlieren observation

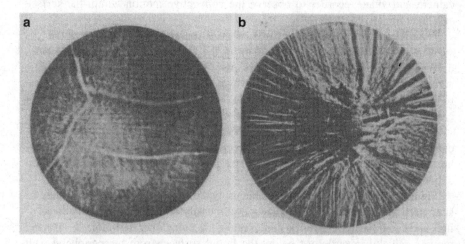

Fig. 4.6 Schlieren pictures of convection lines in surface. (**a**) Low evaporation rate and (**b**) high evaporation rate

through to the centre where they disappear. From features in the lines, their inward radial velocities are in the range 1–10 mm/s.

Combining these findings with the intermittent cold and hot temperature spikes discovered with the fixed micro-thermometers, it is concluded that the intermittent convection is co-operative in nature. Indeed, the Schlieren pictures show patterns of convection lines which are similar to those observed during extensive studies of the surface evaporation of water by Rayleigh-Bénard convection [14, 15].

In conventional Bénard convection [14], the liquid surface is stationary in the laboratory frame of reference, and hence the cells are stationary, and are observed as a regular array of hexagonal convection or Bénard cells. In each cell, hot fluid rises in the centre and cold fluid falls along its boundary (see Fig. 4.7).

In contrast, the surface of a cryogenic liquid, to a depth well below the surface sub-layer, has a velocity directed radially inwards, so that the Bénard cells are drawn out in the moving surface layer into wedge shapes with their boundaries moving radially inward as so-called "streamers".

Each spike in temperature in the vertical temperature profile (of Fig. 4.4) is then due to the passage, past the fixed micro-thermometer, of the edge of a Bénard cell (for a cold spike) and the centre of a cell (for a hot spike).

Another important observation from the video Schlieren recordings is that the number of convection lines, and hence the number of cells enclosed by the convection lines, increases with evaporation rate. Indeed, there appears to be a linear variation between evaporation rate and total length of convection lines or streamers in the surface. The evaporative mass flux per meter length of convection line was never specifically measured, but it is estimated from the Schlieren video pictures to be of the order of 7 g/m for LIN over a wide range of surface mass fluxes from 1.0 to 20 g/m² s.

If one could devise an experiment to scan the surface and measure the local instantaneous evaporation mass flux, one should see a correlation between the

Fig. 4.7 Laboratory simulation in water of Rayleigh-Bénard convection. After Koschmieder and Pallas (1974)

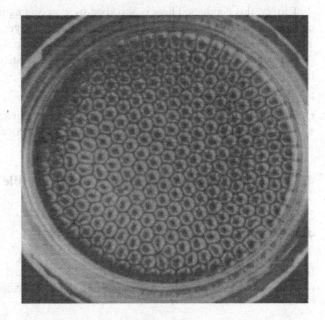

passage of a streamer (the edge of a Bénard cell) and a dip in local evaporation mass flux and surface temperature (at the top of region 2).

Since the streamers are transient in their existence, and variable in number and length with time, it follows that the observed evaporation rate is variable with time, when averaged over integration times of the order of a second. This is particularly so at normal (low) storage evaporation rates where the Schlieren pictures show that only a small length of streamer per square meter of surface is apparently needed to drive the evaporation.

4.6.5   The Delicate Evaporation Impedances of the Surface Sub-layer

Turning (4.9) and (4.10) around, we can see that there are three thermal mechanisms which contribute sequentially, by intermittent convection, mixed conduction/convection, and molecular evaporation respectively, to the equilibrium evaporation within a surface sub-layer of about 5 mm thickness.

Firstly, a series of convective Bénard-type cells, with a vertical dimension of about 5 mm, carry superheated liquid up to within about 0.4 mm of the surface through a temperature gradient of the order of 50 K/m. i.e. the convective heat flow is transferred by a strong thermal process with a small impedance.

Secondly, heat is transferred towards the surface, within the 0.4 mm mixed conduction/convection layer, via a very large temperature gradient of the order of 5000–10,000 K/m, by a relatively weak thermal process. With a high thermal impedance, the process consists of a static thermal conductance enhanced about 1.5–2.5 times by penetration of some of the intermittent convection from the Rayleigh/Bénard convection below.

Thirdly, this conducted heat is absorbed by evaporating molecules of the liquid escaping from a thin molecular evaporation region or layer, probably thinner than 1 μm, with a relatively low thermal impedance for a clean surface.

For example, for LIN, and LCH4 respectively, the relative values of the thermal impedance associated with each mechanism are:

- 0.02 and 0.1 for the convection term,
- 250 and 545 for the mixed conduction/convection term and
- 11 and 28 for the molecular evaporation term with α equal to 10^{-3}.

4.7   Surface Sub-layer Agitation and Unstable Evaporation Phenomena

4.7.1   Agitation of the Surface Sub-layer

It can be seen that the large impedance offered by the mixed conduction/convection of the heat flow across region 2, only 0.4 mm thick, is the dominant term.

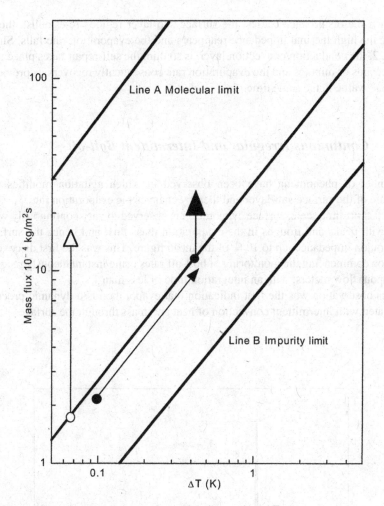

Fig. 4.8 Variation of evaporative mass flux with bulk superheat ΔT.
Line A is upper limit set by molecular evaporation.
Line B is lower limit set by surface impurity.
O-O-Δ depicts equilibrium-equilibrium—Mode 1 rollover mass fluxes in Fig. 5.7.
O-Δ depicts equilibrium—Mode 2 rollover mass fluxes in Fig. 5.8

It is this thin high-impedance region which is separating the bulk superheated liquid from the surface molecular evaporation region, and is preventing a much larger evaporation mass flux from taking place.

Any disturbance or agitation of this thin conduction/convection region, whereby motion of bulk superheated liquid penetrates through it, or replaces it, will result in an immediate, rapid and large, increase in evaporation rate. The surface impedance drops to that of the molecular evaporation impedance and assuming that the evaporation coefficient remains at 10^{-3}, the evaporation rate rises 23 fold for LIN, and some 19 fold for LCH4, as indicated in Fig. 4.8 [15].

When the disturbance ceases, the surface sub-layer regions re-establish themselves, the high thermal impedance reappears and the evaporation rate falls. Since region 2, the conduction/convection layer, is so thin, the self-repair takes place in a few seconds or minutes and the evaporation rate consequently recovers its previous "normal" value in the same time.

4.7.2   Continuous Irregular and Intermittent Boil-off

A number of phenomena have been observed in which agitation modifies the structure of the surface sub-layer and thereby changes the evaporation rate.

On a short time scale, surface evaporation is observed to vary continuously with time, with peaks and troughs in the evaporation mass flux, and hence the surface evaporation impedance, up to 10 % of the mean figure. This was the first discovery made on commencing the monitoring of boil-off rates using instantaneously recording vapour flow meters, with an integration time of less than 1 s.

This observation was the first indication that evaporation is a dynamic process associated with intermittent convection of heat and mass through the surface.

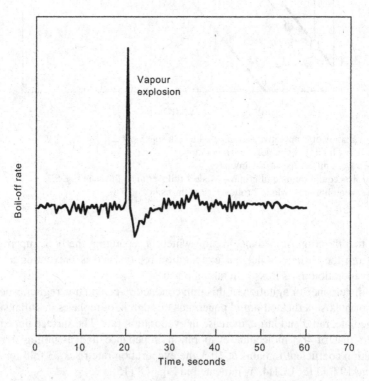

Fig. 4.9 Typical time variation of evaporation rate during a vapour explosion

4.7.3   Vapour Explosions

When the liquid is disturbed by, for example, the transmission of building vibration, or by accidentally knocking the liquid container, the evaporation rate rises rapidly to a high value, and then drops back again just as quickly. The surface evaporation impedance mechanism has broken down quite suddenly and then rapidly repairs itself (see Fig. 4.9).

This evaporation spike can also be reproduced repeatedly by tapping the liquid container at fairly lengthy intervals. Frequently, the flow meter goes off-scale during the spike, but rapidly came back on scale towards the previous reading. Rocking the container so as to cause the liquid to slosh, will also create a boil-off spike.

These boil-off events are called "vapour explosions" and demonstrate how sensitive the evaporation impedance of the surface is to disturbance.

We have found that a demonstration dewar, with a narrow vapour exit into the atmosphere and containing some 5 L of LIN, exhibits "puffs" of vapour, equivalent to mini-vapour explosions, all the time. We have frequently demonstrated this phenomenon in lectures, and it works most obligingly to confirm the irregular surface evaporation of cryogenic liquids.

Vapour explosions are, of course, alarming, but they are surface evaporation phenomena and are generally not dangerous when they arise from transient disturbances of the surface layers. However, because LNG has a low viscosity, sea tankers experience violent sloshing in heavy seas, leading to many vapour explosions along the tank walls. These localised explosions may be the reason for sloshing damage reported in some membrane tanks.

4.7.4   Rollover and Nucleate Boiling Hot Spots

It is possible for agitation of the surface sub-layer to be maintained continuously over a longer period of time, several seconds, or minutes, or hours in length, so as to prevent the self-repairing mechanism re-establishing the equilibrium sub-layer structure.

The high evaporation rate will then be maintained until the thermal overfill energy, or superheat, of the whole volume of bulk liquid is dissipated by the latent heat of the evaporated mass flow.

Large diameter vents are needed to remove the large volumes of vapour produced and to avoid damage arising from the high pressures that would otherwise be generated.

Examples of continuous agitation of the surface sub-layer include:

(a) The spontaneous mixing by double diffusive convection between stratified layers of different composition and temperature, called rollover. The rollover mixing event occurs when the characteristic oscillating, penetrative convection loops approach the surface sub-layer in Mode 1 rollover, and break through the surface sub-layer to the liquid surface in Mode 2 rollover.

(b) Infrequent nucleate boiling from local hot spots whereby a stream of bubbles rise to the surface to disturb and break up the surface sub-layer.

Rollover, as the inevitable consequence of stratification, is more fully discussed in Chaps. 5 and 6.

4.7.5   QHN Boiling

Vapour explosions should not be confused with QHN boiling.

In brief, vapour explosions arise from agitation of the surface sub-layer, temporarily reducing the impedance to the evaporative mass flow.

On the other hand, QHN boiling arises when the whole of the unstable superheated liquid pool or bath becomes full of boiling bubbles, as described above in Sect. 4.4.3, resulting in unpleasant and possibly dangerous venting of cryogenic liquid and vapour mixed together. The only warning of an impending QHN boiling event may be a fall in the BOR below the normal value.

References

1. Collier, J.G.: Convective Boiling and Condensation. Oxford University Press, Oxford (1972)
2. Ashworth, S.P., Beduz, C., Harrison, K., Lavin, T., Pasek, A.D., Scurlock, R.G.: The effect of coating thickness and material on a porous enhanced boiling surface in cryogenic liquids. In: Proc. LTEC 90, 11.2, Southampton (1990)
3. Beduz, C., Scurlock, R.G.: Improvements in boiling heat transfer in cryogenic plant: model of co-operation between industry and university. In: Proc. ICEC12, Southampton, p. 319 (1988)
4. Aitken, W.H., Beduz, C., Scurlock, R.G.: The mismatch between laboratory boiling heat transfer data and industrial requirements. In: Proc. ICEC15, Genoa (1994)
5. Beduz, C., Rebiai, R., Scurlock, R.G.: Evaporation instabilities in cryogenic liquids and the solution of water and CO_2 in liquid nitrogen. In: Proc. ICEC9, Kobe, 802 (1982)
6. Tritton, D.J.: Physical Fluid Dynamics. Van Nostrand, Reinhold (1988)
7. McAdams, W.H.: Heat Transmission. McGraw Hill, New York (1954)
8. Atkinson, M.C.M., Beduz, C., Rebiai, R., Scurlock, R.G.: Heat and evaporation mass transfer correlation at the liquid/vapour interface of cryogenic liquids. In: Proc. ICEC10, Helsinki, 95 (1984)
9. Rebiai, R.: Solubility of non-volatile impurities in cryogenic liquids. PhD thesis, Southampton University (1985)
10. Atkinson, M.C.M.: Cryogenic liquid/vapour and liquid/liquid interfacial mass transfer. PhD thesis, Southampton University (1989)
11. Agbabi, T., Atkinson, M.C.M., Beduz, C., Scurlock R.G.: Convection processes during heat and mass transfer across liquid/vapour interfaces in cryogenic systems. In: Proc. ICEC11, Berlin, p. 627 (1986)
12. Robinson, I.S.: Satellite Oceanography. Ellis Horwood, Chichester (1995)
13. Davies, J.T., Rideal, E.K.: Interfacial Phenomena. Academic, New York (1966)
14. Benard, H.: Ann. Chim. Phys. **23**, 62 (1901)
15. Beduz, C., Scurlock, R.G.: Evaporation mechanisms and instabilities in cryogenic liquids such as LNG. In: Proc. CEC/ICMC, Albuquerque, Adv. Cryog. Eng. vol. **39**, p. 1013 (1994)
16. Koschmieder, E., Pallas, S.: Int. J. Heat Mass Transfer **17**, 991 (1974)

Chapter 5
The Rollover Sequence of Events, Starting with Stratification

Abstract The chapter concentrates on the sequence to be expected if stratification in LNG or LPG cannot be prevented. The natural convective heat flows into the stratified layers will lead to density equilibration between the layers, when spontaneous mixing or roll-over may lead to a large rise in BOR.

The management problems of preventing and/or handling roll-over are discussed in the light of previously recorded roll-over events and simulations. A major problem is the monitoring of all tanks, with limited instrumentation. Small effects like fall off in BOR, and rising liquid level from thermal overfill, in individual tanks may be the only warning of stratification.

Discussion of the spontaneous mixing mechanism illustrates how the violent mixing depends on equilibration of two variables, temperature and composition, at the same time. In other words, roll-over is a property of cryogenic liquid mixtures, not single component cryogenic liquids.

Roll-over is not unique to cryogenics, since oceanographers have met it in stratified sea waters with different temperature and salt content.

5.1 Summary

1. The rollover sequence of events. Confined to tanks at 1 bar with boil-off gas. No rollover with pressurized tanks.
2. Cause and effect relationship between stratification and rollover. The prerequisite of rollover is stratification.
3. The approach to equilibration in density during stratification, between two layers at 1 bar.
4. Convective instability and spontaneous rapid mixing of two layers.
5. Spontaneous mixing by penetrative entrainment of oscillating vertical convection plumes across layer-layer interface disturb surface sub-layer in Mode 1 rollovers, and break through to surface in Mode 2 rollovers.
6. Plumes affect surface evaporation sub-layers causing increased BOR.

© The Author(s) 2016
63
R.G. Scurlock, *Stratification, Rollover and Handling of LNG,*
LPG and Other Cryogenic Liquid Mixtures, SpringerBriefs in Energy,
DOI 10.1007/978-3-319-20696-7_5

7. Release of thermal overfill and heat of mixing during rollover via increased BOR.
8. Mode 1 rollover. Slow build up in BOR with time to ten times normal.
9. Mode 2 rollover. Fast build up in BOR to 100–200 times normal.
10. Two possible mixing mechanisms, but only one in practice.

5.2 Basic Description of "Rollover" and a Stratification Event

The convective instability, reached by the stratification of two multi-component cryogenic liquid layers with different densities, was originally believed to cause the stratified layers to reverse position by rolling over. This is not correct.

What happens in practice is that the densities of the two layers approach each other, for different reasons, over a period of time. As the layers approach density equilibrium, a phenomenon called "double-diffusive convection," involving both thermal and compositional equilibration at the same time, takes over and cause the layers to spontaneously mix.

This mixing, or so-called "rollover", is uncontrolled and may cause a rapid rise in evaporation rate, and tank pressure, associated with the uncontrolled release of the thermal overfill energy from the lower layer, and the heat of compositional mixing of the two layers, within the tank.

If the safety and vent valves on the tank are under-sized, they will not be able to release the large volume of evaporated gas, the tank pressure will rise and may lead to rupture of the tank.

5.3 LNG Rollover: The Sequence of Events

The occurrence of rollover is a hazard, and actions should be taken in advance, to avoid the prerequisite stratification from building up, and to get rid of any stratification when it occurs.

The following sequence of events can be expected.

(1) The creation of density stratification, with small but significant density discontinuities (of the order of 0.5–2.0 %) between layers, by custody management errors, or by unexpected auto-stratification.
(2) The BOR starts to fall below normal BOR, as thermal overfill becomes locked in the bottom layer. The liquid surface in the tank starts to rise, as the bottom layer heats up and expands—only visible in spherical tank filled to 98.5 % or more.
(3) The natural reduction of the density difference between the two layers over time, during normal storage.
(4) Equilibration of the two densities and the uncontrolled rapid spontaneous mixing of the two layers,

(5) The uncontrolled rise in BOR to 5–10 times normal BOR accompanied by steady rise in tank ullage pressure, with Mode 1 rollover.
(6) Relief valves open to relieve pressure rise.
(7) Possible sudden, uncontrolled rise in BOR to a peak of 100–200 times normal BOR within a period of seconds, with Mode 2 rollover. This is the serious concern about an uncontrolled rollover.
(8) Emergency and high speed opening of vents needed to release natural gas vapour and liquid into the atmosphere. There is no warning if Mode 1 self converts into a Mode 2 rollover, and the need for high speed, emergency action to prevent structural damage, should be anticipated.
(9) After the peak in BOR has passed, the subsequent fall in BOR extends over a long period of time before the normal BOR is reached.

5.4   Actions to Avoid Rollover

The actions needed are to try to avoid stratification in the first place, by achieving 100 % mixing of the contents of each tank during all custody management operations, including filling, discharging and long-term storage. Then, rollover will not take place.

However, there are also auto-stratification mechanisms, and if the above mixing is inadequate, then a follow-up mixing procedure is needed urgently to get rid of any residual stratification in every tank.

In addition, the prerequisite of rollover, namely the development of density stratification, needs to be identified in each tank and detected with appropriate instrumentation.

5.5   History of Rollover Events in LNG and LPG Industries

Historically, the LPG industry had, in 1969, recognised the rollover sequence and acted upon the perceived hazard of stratification and excessive vapour production from uncontrolled convective mixing.

The first recorded rollover event with LNG took place at La Spezia, Italy in 1971 [1]. Briefly, a cargo of 33,700 m³ of weathered LNG was bottom-filled into the shore-tank under a lighter heel of 9500 m³. There was minimum mixing during filling and the density difference between the two layers in the tank was about 0.70 %.

Eighteen hours after the filling was completed, an uncontrolled spontaneous mixing or rollover took place. The tank experienced a sudden rise in pressure, and the rollover event was accompanied by the evolution of a huge amount of LNG vapour lasting one and a quarter hours, with the tank vent valves open to the atmosphere (see Fig. 5.1).

Fig. 5.1 La Spezia rollover.
Variation of boil-off rate
BOR with time during
rollover event, showing 250
fold increase in BOR for
1.25 h followed by 2 h
returning to normal BOR

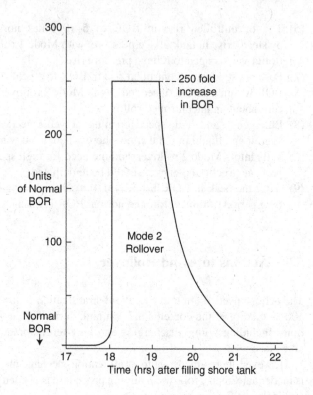

After the vent valves had closed, the tank took another 2 h to reach equilibrium with the normal boil-off rate. Some 500 m³ of LNG were evaporated, as 300,000 m³ of vapour, in the three and a quarter hours of the rollover event, with the peak boil-off estimated to be over 250 times the normal rate, when the rollover converted from Mode 1 to Mode 2. Fortunately the vented gas did not catch fire; neither did the tank suffer damage.

In the next 11 years to 1982, some 40 rollover events were recorded, including four incidents ascribed to nitrogen induced auto-stratification. Since then, development of good practice has reduced the number. The last recorded "near miss" in LNG shipping was in 2008 on a Moss-type 125,000 m³ LNG carrier, when two of the five spherical tanks experienced stratification that nearly resulted in rollovers (see [2], the contents of which have been circulated as a reminder to the whole LNG industry via SIGTTO).

In order to understand the phenomenon of rollover, and how a convective instability can cause such a dramatic event, this chapter will describe and discuss the fluid dynamics of stratification and associated rollover mechanisms. The following chapter, Chap. 6, will discuss the prevention of rollover and, when this fails, the handling of a rollover.

5.6   The Differences Between Single-component and Multi-component Cryogenic Liquids

5.6.1   The T–x Data from Free Boiling Mixtures

To understand rollover, the reader should be aware that most cryogenic liquids are multi-component mixtures and have more complicated evaporation behaviours than a single component liquid. Examples of cryogenic liquid mixtures include:

- Liquid natural gases, LNG, which contain nitrogen, methane, ethane, and smaller quantities of carbon dioxide and higher hydrocarbons.
- Liquefied petroleum gases, LPG, which contain propane, butane and smaller quantities of other hydrocarbons.
- Most cryogenic liquids contain small but finite traces of dissolved impurities such as carbon dioxide, water and hydrocarbons. On exposure to atmospheric air, LNG, LPG and other liquid hydrocarbons at temperatures below 0 °C, may absorb oxygen, nitrogen and carbon dioxide in solution, and water in solution or as a particulate.

It is important, therefore, when storing and handling cryogenic hydrocarbon liquid mixtures, to understand the different evaporation behaviour of cryogenic liquid mixtures compared with that of pure liquids, and how this can affect stratification and the build-up towards rollover.

Thermodynamically, the equilibrium state of a single component fluid is defined by two out of the three variable parameters, pressure P, density ρ, and temperature T, the third parameter being defined by an equation of state.

With a multi-component fluid, an equilibrium state is defined by three or more parameters respectively out of four or more, including P, ρ, T, and x(1), x(2), etc. (where x(1), x(2), etc. are the molar fractions of component (1), (2), etc. in the fluid.) A further complication is that vapour and liquid phases in equilibrium have different compositions.

This is illustrated in Fig. 5.2 by the typical equilibrium (T–x) diagram for a two-component or binary mixture at constant pressure P, where equilibrium is achieved by a freely boiling liquid in contact with its boil-off vapour. "Freely boiling" is the boiling produced by a submerged or wall heater with a sufficiently large heat flux to generate nucleate boiling and multiple streams of bubbles rising through the liquid to the surface.

The difference between vapour and liquid compositions is widely used in fractional distillation for separating a mixture into a lower boiling-component and a higher boiling-point component.

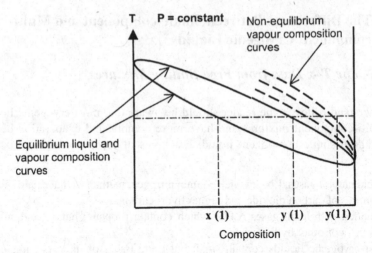

Fig. 5.2 Typical vapour and liquid (T–x) curves under equilibrium (free boiling) and non-equilibrium (surface evaporation) conditions

5.6.2 How LNG Surface Evaporation Yields Much Higher Methane Concentrations in the Vapour

It may not be appreciated that, under storage conditions with surface evaporation only, the well-known (T–x) diagrams, described in Sect. 5.6.1 above, may not apply. The vapour composition y(11), y(22), etc. will not be the same as y(1), y(2), etc. as defined by measuring the liquid composition x(1), x(2), etc. together with T and P, and using the published equilibrium (T–x) data (again, see Fig. 5.2).

The difference arises because, unlike nucleate boiling with vapour bubbles rising rapidly through the bulk liquid and coming into equilibrium according to the (T–x) data, the surface evaporation of each molecular species is differentially controlled by its own diffusion mechanisms, across the surface sub-layer between the bulk liquid and the liquid-vapour interface.

Evaporation-limiting molecular diffusion rates will be determined in a complex manner by the double diffusion of each molecular species within the regions of the surface sub-layer.

In a binary mixture, the different molecular species have different diffusion rates, inversely proportional to their respective molecular weights. The relative concentration of components in the molecular evaporation region of the surface sub-layer, will therefore deviate from the concentrations in the bulk liquid, with an increase in the lower molecular weight component.

The deviation will also increase with increasing heat flow into, and evaporation from, the surface. Consequently, when the heat flows and associated evaporative mass flows increase, the concentration difference between interface and bulk liquid will increase. As a result, the vapour composition y(11) will deviate progressively from the equilibrium value y(1) as the evaporation rate increases, and will contain more of the lower boiling point component (as indicated in Fig. 5.3).

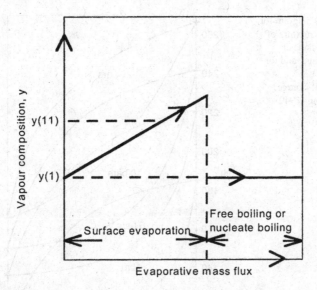

Fig. 5.3 Deviation of vapour composition y(11) from the free boiling value y(1) with increasing evaporative mass flux of liquid mixture

Thus, for an LNG example, as a binary liquid mixture of 80 % methane and 20 % ethane [3], the vapour will possess a 98 % concentration of methane according to the free boiling (T–x) data. However, with surface evaporation only, the vapour composition will increase from 98 % towards 100 % methane from an 80 % liquid, as the evaporation rate increases (see Fig. 5.4).

This change in vapour composition with surface evaporation rate has been studied in experiments at the Institute of Cryogenics, Southampton, after the effect had first been observed.

The consequences of surface evaporation (T–x) values being different to published free-boiling, equilibrium values, means that vapour compositions of LNG and LPG in storage tanks are not a good guide to their liquid phase compositions. See also the consequences of Marangoni film flows up the tank walls in Sect. 6.4.4.

5.6.3 Low Solubility Impurities in the Range 1–10–100 ppm

Until now, this chapter has been largely concerned with totally miscible components in LNG and LPG liquid mixtures. However, there are many common substances which dissolve in cryogenic liquids up to relatively low limits of solubility in the 1–10–100 ppm range.

It is quite clear that, provided the concentration in the solution phase of these minority substances, or impurities, remains well below their solubility limit, there is no problem. Indeed, their presence will probably go unnoticed. Provided there

Fig. 5.4 (T–x) diagram for LNG as binary mixture of methane and ethane at 1 bar bulk storage, and for 5, 10, 20 bar VI pressurised fuel storage. After Ruhemann (1949)

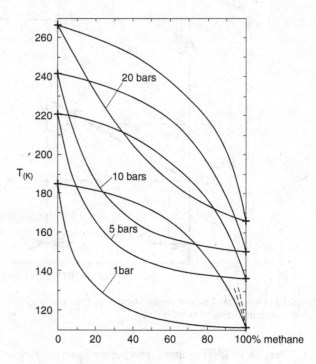

is adequate flushing of the impurity-laden cryogen through the system, then an approach to the solubility limit never happens and all is well.

However, when the cryogen is evaporating in a storage tank, or as part of an industrial process, and there is inadequate flushing, then a mechanism exists whereby the impurity concentration can build up and exceed the solubility limit. The impurity will then pass out of the solution phase and form a solid, firstly as microcrystals in the liquid, and then as a solid deposit.

A number of impurity solubilities have been studied as a function of temperature, using FTIR spectroscopy and gravimetric measurements, and they all vary in the same way via the equation $\log S = A - B/T$. The solubility decreases rapidly and logarithmically with decreasing temperature. If the liquid with dissolved impurity is cooled, the solubility limit may be exceeded very rapidly as the temperature falls and the impurity passes out of solution, possibly depositing on the cold source.

Alternatively, if a liquid solution is expanded is enthalpically through a valve to a lower pressure and temperature, the solubility limit at the lower temperature may well be exceeded and the impurity will pass out of solution. Initially, this will be as submicroscopic crystals throughout the solvent phase, but thermo-diffusiophoresis in local temperature gradients downstream of the valve will aid mechanical deposition on to the walls of the pipework or vessel, where the impurity will collect. The submicron crystals may be unable to grow because the temperature is too low, and they remain as a finely powdered solid with a very large surface area/volume ratio.

Some of these cold finely powdered solid impurities are benign, e.g. water and CO_2, but some may be pyrophoric or spontaneously combustible.

5.7 Layer Stratification in Single Component Cryogenic Liquid with Boil-off at 1 bar

From Chap. 3, it can be seen how liquid convective circulation can lead to density stratification in single component liquids during boil-off at 1 bar. In the absence of any continuous mechanical mixing of the single component liquid, the intrinsic mechanisms of superheated wall boundary layer flow, incomplete evaporation at the surface, and central downward jet of less superheated liquid, can lead to stratification into two layers. Each layer will be at a more or less uniform but different temperature, with a hotter, less dense layer on top of a colder, more dense layer.

As a guide to the associated temperature and density differences, see Table 5.1 which contains estimates of the changes of density with temperature and pressure along the saturation P–T line for a whole range of cryogens. However, note that the density changes are for guidance only, since they relate to the equilibrium saturation condition and not to the superheated liquid from which evaporation takes place or to isenthalpic changes of liquid state.

Since the temperature difference across the liquid-liquid interface between the two layers is small, of the order of 0.1–1.0 K, the mixing effect of thermally driven molecular diffusion in the absence of any convective motion is relatively small. The associated density difference across the liquid-liquid interface acts so as to suppress local convective mixing and the stratification is therefore extremely stable.

However, the heating of the top layer by the wall boundary layer flow continues, leading to a continuing rise in temperature, or an increase in thickness accompanied by a downward migration of the layer interface, or a combination of the two.

When the density difference between the two layers becomes large enough, of the order of 1.0 % (corresponding to an approximate ΔT of 2.5 K for LNG (from Table 5.1), the superheated wall boundary layer flow in the denser, lower layer suddenly has insufficient buoyancy and inertia to penetrate the liquid-liquid layer interface. The boundary layer flow is trapped in the lower layer and no evaporation can take place to release its superheat. Instead, the wall boundary layer turns over at the interface and its motion and superheat is locked into the lower liquid causing it to heat up instead. When this happens, the boil-off rate will fall as the first indication

Table 5.1 Derivative properties of saturated liquid cryogens at NBPs

	NBP (K)	ρ (kg/m³)	$-(d\rho/dT)_{sat}/$ ρ (%/K)	$+(dT/dP)_{sat}$ (K/bar)	$-(d\rho/dP)_{sat}/$ ρ (%/bar)
Nitrogen	77.31	806.8	0.57	10.8	6.1
Argon	83.80	1394	0.47	7.1	3.3
Oxygen	90.19	1141	0.43	12.26	5.2
Methane	111.67	422.4	0.40	14.3	5.7
Ethane	184.55	488.5	0.20	18.1	3.6
Propane	231.1	581	0.19	29.2	5.55

of a stratification effect and associated increase in thermal overfill via the unstable superheated state. As described in Sect. 2.2.3, this type of thermal overfill in a single component liquid may be released by a violent QHN boiling of the lower layer, with consequences such as the ejection of vapour mixed with liquid through the vents, and possible mechanical damage to the storage vessel.

Figure 5.3 illustrates how density stratification leads to the lower layer boundary layer flow failing to penetrate the liquid-liquid interface.

On the other hand, if liquid is stored with zero boil-off under a rising vapour space pressure in a vacuum insulated pressure vessel to a maximum pressure of 10–20 bar or higher, a stable stratification of saturated liquid at the vapour space pressure collects in a layer above a second colder sub-cooled layer. This zero boil-off, pressurised storage is widely used in the cryogenic industry for storing and distributing LIN, LOX, LA etc., and is now being applied for use of LNG as a fuel for trucks and small ships. There appears to be no likelihood of rollover under pressurised storage and handling, but mixing of the fresh LNG with the heel should be achieved before completing refuelling operations.

5.8   Density Stratification in Miscible, Multi-component Cryogenic Liquid Mixture at 1 Bar

With a multi-component cryogenic liquid, the triggering and build-up of stratification can occur in a variety of ways, depending on the identity of the liquid components, and the previous history of the liquid elements.

Once again, the stratification is convectively stable until the densities approach equilibration. Then; mixing across the liquid-liquid interface is controlled by double diffusive convection, with both temperature and concentration gradients enhancing the density-gradient driven, convective mixing.

This stratification in a cryogenic liquid mixture and density equilibration inevitably leads to unstable evaporation, which has acquired the name "rollover". This unstable evaporation takes place when the stratified layers spontaneously mix, leading to a rapid and extended increase in boil-off rate above the normal BOR, which continues until the heat of mixing, and the thermal overfill in the lower layer, have been absorbed by the latent heat of vaporisation in the excess boil-off vapour.

From detailed studies of unstable evaporation from stratified liquid mixtures carried out in Tokyo, MIT, Southampton, and elsewhere [4–10], it has been observed that every experimentally simulated rollover event was spectacular, and different in terms of all measurable parameters, including the time taken to spontaneously rollover, the variation of BOR with time, and the peak evaporation rate generated. No acceptable correlation from the studies was therefore possible.

5.9   The Fluid Dynamic Storage Behaviour at 1 Bar of Two LNG or LPG Layers with Different Density

If two multi-component liquid layers, with different densities and/or different temperatures and a common interface, are allowed to stand in a **thermally isolated** tank, then there is no convection.

Equilibration of composition and temperature takes place relatively slowly through molecular and thermal diffusion across the interface.

When **heat enters** the liquid space through the walls, the heat is convectively absorbed by the upward moving wall boundary layer flow which feeds into the evaporating surface, from which the boil-off gas carries the heat out of the tank. However, when there is a density reduction across the liquid-liquid interface of the two layers (in the upward direction of the stable convection at the wall) the wall boundary flow in the lower liquid layer may have insufficient buoyancy to penetrate this density drop. Instead, the boundary layer flow turns over through 90° at the liquid-liquid interface to become a radial inflow just below the interface and then joins a central downward jet to be mixed into the lower layer. Heat entering the lower liquid via the boundary layer cannot then be released by surface evaporation, and remains trapped in the lower layer (see Fig. 5.5).

Looking analytically at this lack of penetration by the boundary layer flow, the quantity $(d\rho/dT)/\rho$ for LNG and LPG mixtures, is of the order of 0.4 and 0.2 % per K, respectively at their boiling points (see Table 5.1).

Experimental measurements on these mixtures indicate that the local temperature rise in the boundary layer is about 1 K. Thus, the buoyancy or upward force $\Delta\rho g$ on unit element of LNG driving the boundary layer flow is of the order $0.004\rho g$ and this leads to a vertical velocity in the boundary layer of the order 0.1 m/s, as is observed.

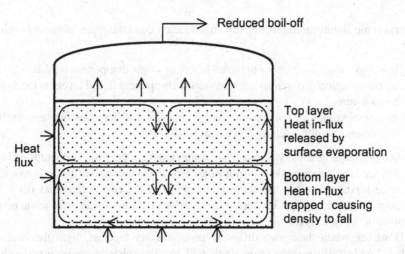

Fig. 5.5 Stratification with low density top layer above higher density bottom layer. Density difference >0.5 %

If the density drop across the liquid-liquid interface is greater than 0.004ρ (or 0.4 % of the mean liquid density) the buoyancy force driving the boundary layer upwards disappears.

However, the momentum of the boundary layer will carry it beyond the interface a vertical distance of a few cm, before the boundary flow turns over.

This is what is observed experimentally; viz. the boundary layer flow turns through 90° away from the wall to join a radial inward flow while its vertical velocity component oscillates about the interface with an exponential decay and an initial amplitude of the order of 0.01 m.

We now have a convective mechanism whereby heat entering the lower liquid layer through the walls and floor of the storage tank cannot be released by the normal process of surface evaporation. Instead, this heat is mixed convectively throughout the lower layer, via the central down ward jet, as thermal overfill. The mean temperature of the lower layer rises, and the mean density falls with time.

At the same time, the density of the upper layer is increasing with time through evaporation of methane from the upper layer. This evaporation, as described in Sect. 5.6.2 above, causes the ethane concentration to increase and the density to rise with time.

Eventually, the layer densities become equal and a convective instability is reached.

Mixing models are able to describe the approach to this convective instability [11], with reasonable precision, but not what happens next.

A different model is needed (see below).

5.10   Double Diffusive Convection and Rapid Spontaneous Convective Mixing of Two Layers of LNG on Approach to Density Equilibration

In cryogenic liquid mixtures, density stratification can take place in the following ways:

(1) by temperature differences between layers in single component liquids,
(2) by composition differences between multi-component liquid layers at the same temperature,
(3) by combinations of temperature and composition differences between multi-component liquids, such as LNGs and LPGs.

A step-change in density across a horizontal interface between two stratified layers will tend to disappear slowly with time through (a) thermal diffusion across the interface tending towards the same temperature in the two layers, and (b) self-diffusion of molecular species across the interface tending towards the same composition in the two layers.

However, when these two diffusion processes act together, by a mechanism called double diffuse convection, they will greatly enhance convection motion across the interface of the two layers. Since convective mixing is very much faster than diffusive mixing, it is the cross effect of both thermal and compositional diffusion processes which enhances the natural convection processes. As the densities approach equilibration, the convective motion and associated mixing starts to dominate what is taking place at the layer-layer interface. The subsequent convective motion increases rapidly, the associated heat of mixing and release of thermal overfill causing the rate of evaporation to rise. The enhanced mixing is uncontrolled once it commences, and is accompanied by a rapid and unexpected rise in BOR. This phenomenon is called "rollover" and the mixing behaviour is a basic property of all cryogenic liquid mixtures, not just LNG.

In particular, this property has serious consequences which increase in magnitude with the rapid growth of the LNG industry, and which need adequate attention in understanding and preventing the occurrence of rollover.

The effects of "double diffusive convection" are not just confined to cryogenic liquid hydrocarbons. They have been widely studied and modelled from the observation of local convective motions in the sea by oceanographers, where heat and salt content are the two diffusion contributors. The standard analysis is given in, for example, references by Turner [12, 13] and by Huppert [14].

5.11   Penetrative, Oscillating Convection Across Layer-Layer Interface, Leading to Uncontrolled Rise in BOR, During a Rollover Event

In fluid dynamical terms, as the densities of two cryogenic liquid layers approach the same value with time, there comes a point at which double diffusive convection manifests itself in the following remarkable way, which has been observed about 100 times experimentally, and also recorded by video camera [15].

Since double diffusive convection is not unique to LNG or any other cryogenic liquid mixture, experimental rollover simulations have been studied with LA/LIN, LOX/LIN, and liquid Freon mixtures, as well as LNG mixtures.

In all cases, the liquid-liquid interface is observed, during video recordings of flow visualisations of real rollovers, to become unstable whereby upwards penetrative oscillations of the interface build up in vertical amplitude. Elements of the lower layer are mixed into the upper layer, and the liquid-liquid interface migrates slowly downwards.

At the same time, the evaporation rate starts to rise, and goes on rising as the convective mixing oscillations slowly grow in amplitude and start to come into contact with the surface sub-layers. This relatively slow build-up in BOR is called a "Mode 1 Rollover".

Figure 5.6 shows a sequence of flow visualisation photographs, of an experimental rollover between stratified layers of Freon 11/113 mixtures. These illustrate the downward migration of the interface over a period of 172 min followed by its rapid disappearance after total mixing by penetrative convective oscillations after 182 min [7].

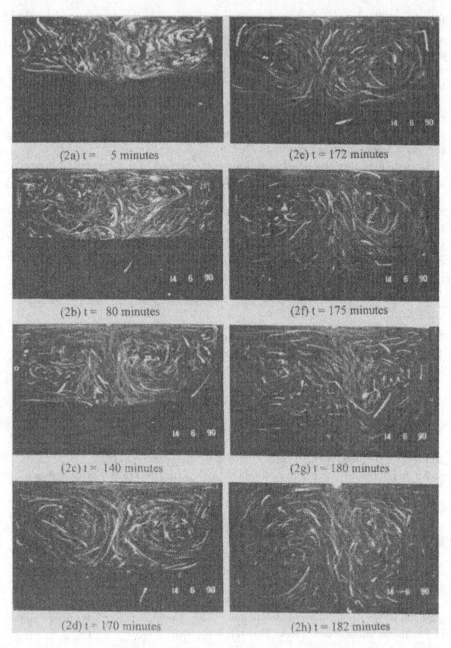

Fig. 5.6 Sequence of flow visualisation photographs showing spontaneous mixing (rollover) of two stratified layers. Exposure time was 4 s to show trajectories of seed particles

In some of the rollovers, when the amplitude of the convective penetrative oscillations increased so that the oscillations penetrate through the surface sub-layers to the liquid surface, the BOR rose very rapidly by 10–50 fold. This behaviour is called "Mode 2 Rollover" in which the convective oscillations are believed to break up the morphology of the surface sub-layer, particularly the high impedance thermal conduction region. This results in the unimpeded and much higher molecular rate of surface evaporation to be realised (see Sect. 4.6.2 on molecular evaporation rates).

This high evaporation rate is then maintained so long as the convective oscillations continue to disturb the surface — up to the time when the two layer interface disappears and a homogeneous density mixture is produced. Then the rate drops to an intermediate high value corresponding to the excess superheat temperature across the "new" surface sub-layers of the now self-mixed (by convection) liquid of homogeneous density, but not yet homogeneous in temperature and composition. Subsequent residual (compositional and thermal) mixing towards homogeneous temperature and composition is then accompanied by the intermediate high evaporation rate after the peak BOR, before the final mixture, homogeneous in temperature, composition and density, is arrived at naturally.

Optical observations of the transition through the critical point in a critical point cell (when the density of vapour and liquid phases equalise) show similar oscillatory, penetrative convection across the phase boundary.

This model, whereby penetrative convective oscillations reach the surface and break up the surface sub-layer, enables the magnitude of the very high peak boil-off rates to be predicted, for Mode 2 Rollover.

While the estimated evaporation coefficient for LIN is 10^{-3}, the value for LNG is not known; it could be larger by a factor of 10, thereby allowing the peak boil-off rate to rise to 190 times the normal BOR, close to that observed in the La Spezia incident.

5.12   Release of Thermal Overfill During Rollover

In terms of thermal overfill, we have seen how the heat flow entering the lower liquid layer by thermal conduction from ambient through the insulation is trapped in the lower layer, causing its temperature and energy content (superheat or thermal overfill) to rise with time. The build up of thermal overfill within the lower layer is clearly triggered if, or when, the density difference between the two layers exceeds some critical value, which appears to be the order of 0.5 % (0.70 % for the La Spezia event).

The occurrence of two stratified layers is then the forerunner of the sequence of mechanisms leading to rollover, described in Sect. 5.3.

When rollover takes place, it is the thermal overfill energy in the tank, not just the lower layer, and the compositional heat of mixing between the two layers, which is released via the latent heat of the venting vapour. The maximum rate of generation of vapour is determined by the molecular surface evaporation limit—it is not infinitely great like an explosion.

From the 100+ instrumented, experimental rollover events made at Southampton, the following conclusions can be made:

• There was never any nucleate boiling—only surface evaporation.
• The envelope of the BOR-time graph is different for every rollover.
• The total additional evaporated mass equates to the total thermal overfill energy, plus the compositional heat of mixing of the two layers, released during the rollover.
• Once a stratification of less dense layer on top of a more dense layer is set up, and heat energy is supplied to the lower layer, then this energy will only be released spontaneously by an inevitable rollover event.
• The peak evaporation rate may be the order of 50–250 times the normal storage rate, but it is not infinite.

5.13 Experimental Studies: The Observation of Two Modes or Types of Rollover

Experimental observations on laboratory generated rollovers have been made by several groups, including those at Southampton using video recording of the flow visualisations during the rollover mixing of previously stratified layers. These studies showed that there is a whole spectrum of variations of evaporation rate with time between the extremes of two modes of rollover. These modes were identified by dramatically different evaporation behaviours, and by somewhat different rates of migration of the liquid-liquid interface separating the layers.

Figures 5.7 and 5.8 show summaries of two experimental runs following the setting up of two stratified layers, which demonstrated Mode 1 and Mode 2 rollovers respectively. The difference is very significant.

In the first case, Mode 1, the BOR increased relatively slowly with time to a peak value over a period of 60 min or more, and then subsided equally slowly back to the normal storage value. The excess evaporation tracked the rise and fall of liquid superheat in the upper layer, as the mixing proceeded spontaneously to completion and as the thermal overfill was released. The interface also moved down very slowly. All was very peaceful but uncontrollable, without any emergency venting to the atmosphere.

In the second case, Mode 2, after 350 min, the BOR rose very sharply within a few seconds to a high peak value, at least 20 times the normal BOR, before subsiding. At the same time, the interface moved downward much more rapidly than in Mode 1.

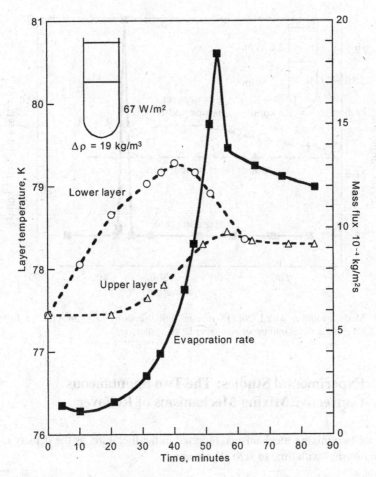

Fig. 5.7 Mode 1 rollover with LIN/LOX mixture. Initial density difference 19 kg/m³ (2.5 %). High heat flux of 67 W/m² into lower layer only

Mode 2 behaviour had features similar to that of a vapour explosion described in Chap. 3.

The rollover studies at Southampton included the use of binary mixtures of nitrogen and oxygen in two stratified layers which enabled the measurement of vertical temperature profiles, vertical composition profiles and evaporation rates, all as a function of time up to and after the inevitable rollover.

The studies concluded that all binary cryogenic liquid mixtures in two stratified layers, at 1 bar pressure, will spontaneously mix or rollover with an uncontrolled increase in BOR, including LNG, LPG, LIN/LOX, LIN/LA and Freons. However, the studies were unable to pinpoint any measurable factors which might help to distinguish whether the inevitable rollover would be Mode 1 or Mode 2.

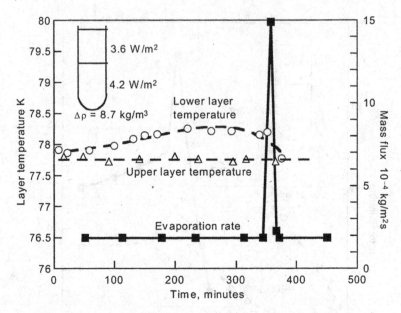

Fig. 5.8 Mode 2 rollover with LIN/LOX mixture. Initial density 8.7 kg/m³ (1.0 %). Low heat fluxes of 3.6 and 4.2 W/m² into upper and lower layers respectively

5.14   Experimental Studies: The Two Spontaneous Convective Mixing Mechanisms of Rollover

There are two mixing mechanisms, reported in the literature, as the density difference diminishes with time to zero. These are:

(a) entrainment by wall boundary layer flow penetration, and
(b) entrainment via vertically oscillating convective plumes, simultaneously across the whole of the liquid-liquid interface.

Both mechanisms cause oscillating upward entrainment mixing of the bottom layer into the top layer, and downward motion of the interface.

From the laboratory rollover simulations and numerical modelling at Southampton [4, 7], both wall-heating and base-heating of the bottom layer generally led to entrainment mixing by oscillating convective plumes only. The oscillating motion of the convective plumes, upwards into the top layer and downwards into the bottom layer, increased in amplitude with time as the spontaneous rollover proceeded; while the average position of the interface moved slowly down.

When the amplitude of the oscillating convective plumes built up slowly from small oscillations and never reached the top layer, then Mode 1 rollover behaviour was observed both in the experimental simulations and in the numerical and theoretical modelling (see Fig. 5.9).

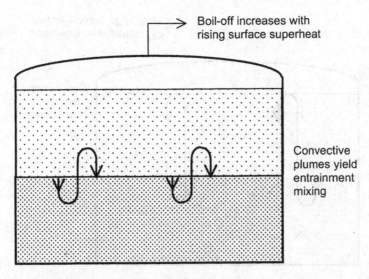

Fig. 5.9 Mode 1 rollover with plumes penetrating upper layer

With slight wall-heating only, it was possible to induce wall boundary flow penetration and mixing, but this was a special case. Mixing by convective plumes was, without doubt, the dominant mechanism in the experimental rollovers.

For Mode 2 events, the amplitude of the oscillating plumes was observed to increase very quickly and almost immediately penetrate through the surface sub-layer structure into the molecular evaporating region at the liquid-vapour interface.

This penetration appeared to disturb the delicate morphology of the whole of the surface layers, so that superheated liquid from the bottom layer replaced the surface evaporating layer, enabling the evaporation rate to rise 20–50 fold, i.e., to the molecular limit, just like a vapour explosion (see Fig. 5.10).

Because this mixing was so intense across the whole liquid surface, the interface moved quickly downwards at the same time as the evaporation rate rose to a peak. Once the mixing was complete, the convective plumes disappeared, the disturbance of the surface layers stopped, and the evaporation rate fell relatively quickly to the intermediate value determined by the liquid superheat, as the surface sub-layer repaired itself into its normal morphology.

In some experimental rollovers, Mode 1 behaviour converted into Mode 2 when the penetrative oscillating convective plumes increased in amplitude with time and penetrated the evaporating surface sub-layer.

One feature of the numerical modelling was that the intensity of the mixing was influenced by the ratio of total base heat flux to total wall heat flux. Generally, a higher heat flux ratio, greater than unity, led to more intense final mixing. It should be noted that, in the La Spezia LNG incident, the ratio of base to wall heat flux was high at 3.0. It may therefore be concluded that the peak boil-off during a rollover is significantly reduced if the base heat flux is no larger than the wall heat flux.

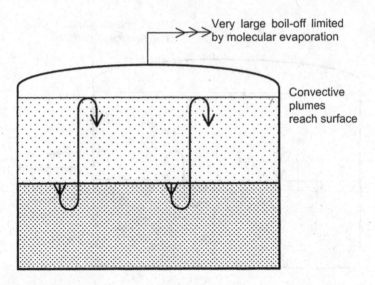

Fig. 5.10 Mode 2 rollover, with plumes breaking into surface sub-layer

It should be additionally noted from the experiments that:

- all stratifications of two layers led to naturally occurring, spontaneous rollovers,
- the peak evaporation mass fluxes observed during Mode 2 rollovers, and some Mode 1 rollovers, could, on scaling up to storage tank dimensions, exceed the venting capacity of the installed safety valves, and lead to structural damage.

During all the experimental studies at Southampton, at no time did the layers mix or exchange position by "rolling over". In every case, the mixing occurred via oscillating penetrative convective plumes across the whole of the liquid-liquid interface separating the two layers.

The term "rollover" is therefore not a correct description of the spontaneous mixing, but since the term is in common usage to describe a single dramatic event, we shall continue to use the term also.

References

1. Sarsten, J.A.: LNG stratification and roll-over. Pipelines Gas J. **199**, 37 (1972)
2. K-Line Ship Management Co. Ltd, Tokyo. Density-stratified liquid layers and rollover in Moss Type tank LNG carrier "DEWA MARU" (2008)
3. Ruhemann, M.: The Separation of Gases. Oxford University Press, Oxford (1949)
4. Agbabi, T., Beduz, C., Scurlock, R.G., Shi, J.Q.: Evaporation stability of cryogenic liquids under storage. In: Proc. LTEC 90, 1.5 (1990)
5. Shi, J.Q., Beduz, C., Scurlock, R.G.: Numerical modelling and flow visualisation of mixing of stratified layers and rollover in LNG. Cryogenics **33**, 1116 (1993)

6. Scurlock, R.G.: Stability of cryogenic liquids under storage. In: Proc. Kryogenika 90, Kosice (1990)
7. Agbabi, T.: Rollover and interfacial studies of LNG mixtures. PhD thesis, Southampton University (1987)
8. Shi, J.Q. Numerical modelling and experimental study of rollover. PhD thesis, Southampton University (1990)
9. Sugawara, Y., Kubota, A., Muraki, S.: Rollover test in LNG storage tank and simulation model. Adv. Cryog. Eng. **29**, 805 (1983)
10. Smith, K.A., Lewis, J.P., Randall, G.A., Meldon, J.H.: Mixing and rollover in LNG storage tanks. Adv. Cryog. Eng. **20**, 124 (1974)
11. Germales, A.E.: A model of LNG roll-over. Adv. Cryog. Eng. **21**, 330 (1975)
12. Turner, J.S.: Buoyancy Effects in Fluids. Cambridge University Press, Cambridge (1979)
13. Turner, J.S.: The complex turbulent transport of salt and heat across a sharp density interface. Int. J. Heat Mass Transfer **8**, 759 (1965)
14. Huppert, H.E.: On the stability of a series of double-diffusive layers. Deep Sea Res. **1005** (1971)
15. Beduz, C., Scurlock, R.G.: Spontaneous convective mixing or "roll-over" between two stratified layers of LIN/LOX mixture. Video tape demonstrated at Heat Transfer Conference, San Francisco (1986)

Chapter 6
Factors Creating Stratification: Management of LNG Rollover

Abstract Stratification or the creation of two convectively stable layers of liquids with different densities is the beginning of events leading to possible roll-over.

The main factors are the convective mechanism whereby heat inflows into the lower layer become locked into that layer causing its temperature to rise and density to fall with time; also the preferential evaporation of methane from the upper layer of LNG (or propane from stratified LPG) will lead to an increase in density with time; also the heat of mixing is large and may evaporate up to 10 % of the liquid mixtures.

The chapter discusses how stratification can be prevented, and/or removed, by adequate mixing operations. However, auto-stratification after filling is possible and needs to be countered by mixing at a later time.

The chapter concludes with some advisory points towards anticipating roll-over, if the mixing is insufficient, or too late, to remove stratification.

6.1 Summary

1. There are several ways in which LNG, a multi-component mixture, can become stratified in a storage tank.
2. Filling, refilling or re-fuelling with a different density LNG to that already in the tank, causing density stratification.
3. Auto-stratification mechanisms.
4. Stable stratification in pressure tanks without boil-off loss, and no rollover.
5. Filling with thermally sub-cooled LNG or LPG is a dangerous practice to be avoided.
6. Path dependent mixing and variable, vapour flash volumes.
7. Possible internal mixing devices.
8. Handling a rollover.

© The Author(s) 2016   85
R.G. Scurlock, *Stratification, Rollover and Handling of LNG,*
LPG and Other Cryogenic Liquid Mixtures, SpringerBriefs in Energy,
DOI 10.1007/978-3-319-20696-7_6

6.2 Density Stratification

As described in Chap. 5, the density difference which can lead to rollover is as small as 0.5 %. This small size of the requisite for rollover has been seen experimentally on many rollover events. Indeed, the first reported event, the La Spezia rollover, was initiated by a density difference of only 0.7 %.

This small difference is difficult to detect directly with available in-tank instrumentation of multiple temperature measurements, densitometers or BOR monitoring. Furthermore, the success of any mixing operation to remove stratification is difficult to judge. So, how do we go forward?

The first step is to prevent stratification in the first place [1], while the second step is to have effective mixing capability in each tank as the simple answer for avoiding a rollover. The third step is to have adequate in-tank instrumentation. Failing these steps, then eventually you may be faced with handling an uncontrolled rollover.

6.3 Custody Management Creating Two Layers of LNG

6.3.1 The Standard Loading Mistake

The standard loading mistake is to fill a tank with a layer of methane-rich, low density liquid on top of a lower layer or "heel" with a higher density, without mixing. The two layers are convectively stable with no mixing and everything appears to be under control, but not for long, because this convective stability will not last. The first recorded rollover event with LNG was met in La Spezia in 1971 following this simple loading mistake. So let us consider the consequences of this mistake in an example, assuming LNG is a two component or binary mixture of methane and ethane.

Firstly, let us consider a tank filled with uniform density LNG at its normal boiling point producing boil-off gas and with no stratification present. Because it is a mixture, there is a difference in composition between vapour and liquid so that the vapour is richer in the lower boiling point component—methane, normal boiling point 112 K, liquid density 424 kg/m^3—leaving the liquid richer in the higher boiling point component—ethane, normal boiling point 184 K, liquid density 544 kg/m^3. It follows that, as surface evaporation continues under normal, well-insulated, storage conditions, the originally well mixed liquid will become progressively richer in ethane, and the density and boiling temperature, at constant tank pressure, of the liquid mixture will increase steadily with time.

Let us now look more closely at an element of superheated liquid mixture at the liquid/vapour interface during surface evaporation. The spent liquid, after evaporating a methane rich vapour, has increased in density due to both evaporative latent heat of cooling and an increase in more dense ethane composition. The spent element is convectively unstable and sinks away from the surface.

Together with all the similarly spent elements, it mixes convectively with the bulk liquid producing a uniform density and composition. There is no stratification.

Secondly, let us now consider a fresh supply of the same original LNG binary mixture being is added to the "old" or "weathered" liquid, or "heel", without being mixed with it. The fresh liquid will have a lower density and lower temperature and will tend to sit on top of the more dense heel in a convectively stable condition. Thus a stratified top layer has been accidentally, or unknowingly, established, and this density stratification is convectively stable to start with, but again not for long.

If nothing is done to get rid of this density stratification, the sequence of events described in Sect. 5.3 will surely take place and an uncontrolled rollover may happen.

6.4   Auto-stratification Specific to LNG and LPG Mixtures

There are a number of scenarios whereby auto-stratification may occur, including the following.

6.4.1   Mechanisms Due to Density Differences

Since stratification is the necessary precursor of a rollover event, it is important to consider the various ways in which stratification can occur, can be anticipated, and can be prevented.

We have already noted that density differences between LNG or LPG layers can occur from combinations of temperature differences and composition differences. Also, that wall boundary layer flow cannot penetrate a density difference greater than about 0.5 % between the layers, in which case the heated wall boundary layer flow is trapped in the bottom layer.

Table 5.1 shows that a 1 % density difference corresponds to a temperature difference of about 2.5 K in liquid methane and 5.0 K in liquid propane at their respective normal boiling points.

Likewise at constant temperature, a 1 % density difference corresponds to a 5 % change in methane concentration in LNG as a binary mixture of methane and ethane, and a 15 % change in propane concentration in LPG.

These figures suggest that auto-stratification can be expected to be a common occurrence, since the necessary density changes are so small.

There are several examples of auto-stratification particular to cryogenic mixtures of LNG.

6.4.2   High MW Volatile Component in the Mixture

During surface evaporation of a mixture, the more volatile, lower boiling point component evaporates preferentially, leaving a higher concentration of the less volatile, higher boiling point components in the "spent" surface layer. If the evaporating component has a density (or molecular weight, MW) greater than the higher boiling point components remaining behind, then the "spent" layer has a reducing density as surface evaporation proceeds and will spontaneously grow into an upper stratified layer, or auto-stratify.

An important example is LNG with a significant amount (>1.0 %) of nitrogen.

During storage of LNG with, say, a composition containing 3 % nitrogen, preferential evaporation of the more volatile nitrogen with $MW = 28$, liquid density 808 kg/m^3, from LNG with $MW \sim 16$, liquid density ~ 430 kg/m^3, will result in a liquid density reduction towards 420 kg/m^3 of 2.5 %, which is more than enough for auto-stratification.

It should be noted that the enhanced composition of the vapour with the lower boiling point constituent, due to the surface evaporation effect discussed in Sect. 5.6.2, will further enhance the density reduction in the top layer required for auto-stratification to take place.

6.4.3   Non-volatile Impurities in the Surface

If carbon dioxide, water, or any other non-volatiles are present in solution, which may be at concentrations of 1 ppm. or considerably less, then during surface evaporation, they do not evaporate and may subsequently remain at the surface as a monomolecular film. This film may be continuous, or it may consist of discontinuous floating rafts of impurity; but when it forms, it will provide a significant additional impedance to the surface evaporation process and literally switch off the evaporative mass flow.

Thus, we have another triggering mechanism whereby the superheated liquid from the wall boundary flow cannot evaporate and a superheated layer forms immediately below the surface.

Non-volatile films may also arise from condensation of carbon dioxide, water, etc. entering the ullage vapour space during custody management operations, or from leakage of atmospheric air.

Laboratory studies with freely evaporating cryogenic liquids have witnessed on many occasions the switching off of evaporative mass flows for periods of many minutes, and the subsequent rapid rise in evaporation, or vapour explosions, beyond the upper limit of the flow meter. However, the introduction of carbon dioxide, either into solution, or into the ullage space, in order to stimulate, or simulate, the formation of a film and switch off the evaporation, proved to be inconclusive.

These monomolecular films do, of course, exist in ambient temperature environments and can be used to reduce the evaporation of water stored in reservoirs. They are also responsible for the stability of smog particles in the atmosphere.

The Schleiren photographs of evaporating cryogenic liquid surfaces frequently show areas where there appears to be no evaporation.

6.4.4   Recondensers Producing Lower Density Condensate

Another source of auto-stratification arises when the boil-off gas is recondensed and returned to the same tank. Generally, the boil-off gas contains a higher proportion of the lowest MW components and condenses to a liquid with a lower density than the

parent liquid in the tank. If the condensate is returned without mixing, then it will tend
to collect as a top layer above the parent liquid; a stratification has been created. The
answer, of course, is to return the lighter condensate to the bottom of the tank to ensure
adequate mixing. This precaution applies currently to LPGs where they are being
recondensed: also possibly to LNGs, if recondensers are introduced in the future.

6.4.5   *Marangoni Film Flow Effect*

It may not be commonly known that the tank wall above the liquid level in a LNG
tank is wetted by surface tension driven film flows up the tank wall. This effect, an
example of the so-called Marangoni effect, was first studied at Southampton in
1974 [2]. Have a look to see for yourself!! Then compare what you see with the
tear-drops round the edge of a glass of sherry or port (see Fig. 6.1).

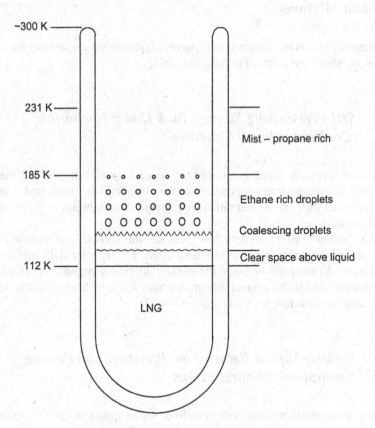

Fig. 6.1 Marangoni effect. Schematic diagram showing appearance of transparent container wall,
with film flows above LNG, together with observed temperatures

Due to surface tension differences between methane and ethane, a film of ethane-rich liquid is drawn up the wall, losing methane by evaporation, until pure ethane droplets build up at the 185 K (NBP of ethane) temperature level. The ethane droplets run back down the wall, and being more dense than LNG, they collect over a long period of time at the tank bottom as a thin, warm, dense, stratified layer. Above the 185 K level, a film of propane rich liquid is drawn up the warmer levels of the wall and propane droplets appear at the 231 K level (NBP of propane) if there is sufficient concentration in the LNG (Fig. 5.9).

However, measurements of the mass dynamics of the surface-tension driven separation process show that the mass flows in the films are not great enough to be commercially viable for extracting ethane or propane from LNG [3].

Another point to watch is that the wetting of walls and cold exit piping, due to the Marangoni effect, make the monitoring of vapour composition particularly difficult.

6.5   Auto-stratification in Both Single Component Liquids and Mixtures

Auto-stratification is not confined to cryogenic liquid mixtures, and can occur in a number of other ways relevant to handling LNG.

6.5.1   Self-Pressurising Storage Tank Under Isochoric (Constant Volume) Conditions

If a tank of cryogenic liquid is allowed to pressurise, with zero boil-off, then the evaporation is progressively suppressed as the ullage pressure rises, and a surface layer is created with a rising saturation temperature in equilibrium with the increasing ullage pressure.

The superheated wall boundary flow feeds into this surface layer without evaporating, and a saturated upper layer builds up above a lower layer at the initial filling temperature. As the pressure rises, the density difference between the two layers remains small, so that the auto-stratification is very stable with no sudden spontaneous mixing and therefore no rollover (see Chap. 7).

6.5.2   Variable Vapour Return Line Pressures and Passing Atmospheric Weather Fronts

A falling atmospheric pressure with time, from say an approaching storm front, will cause the BOR to rise from a freely venting tank, and the bulk temperature to fall by up to about 1 K, until the storm front passes.

A rising atmospheric pressure with time after, say, the passing of a storm front, will tend to suppress evaporation and lead to the mechanism of stratification outlined in Sect. 5.7 above. The surface layer temperature will rise to maintain equilibrium with the rising atmospheric pressure, while the bulk liquid temperature will remain constant. The associated density difference between the two layers, even for a severe storm with a pressure drop up to 100 mbar, is unlikely to exceed 0.5 % for most LNG liquids. The trapping and switching of wall boundary flow into the lower bulk liquid appears therefore to be unlikely, during severe storms.

Pressure variations in the vapour return lines will also produce temperature and density variations in a similar manner. If the pressure rises by more than 100 mbar, then the triggering of auto-stratification may be expected.

With a brim full tank, a pressure drop may lead to liquid expansion and liquid being carried into the vapour line, with uncontrolled consequences.

6.6 Custody Management Filling with Thermally Subcooled Liquid Creating Thermal Underfill: A Dangerous Situation

If a tank is topped up with thermally subcooled liquid—a dangerous practice—then auto-stratification will take place immediately (see Fig. 6.2).

Thermally subcooled liquid, i.e. liquid cooled to below its boiling point at tank pressure, will be more dense than the heel of old liquid. If mixing is incomplete, the old liquid will rise above the in-coming subcooled liquid and collect as a stratified upper layer.

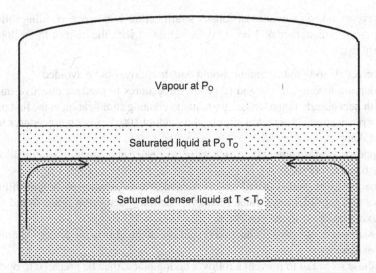

Vapour at P_O

Saturated liquid at $P_O T_O$

Saturated denser liquid at $T < T_O$

Fig. 6.2 Autostratification by addition of sub-cooled liquid. For LNG, when density difference between layers >1.0 %, or sub-cooling of lower layer >3 K. For LPG, sub-cooling of lower layer >5 K. This will lead to (**a**) sub-atmospheric pressure in ullage space and (**b**) rollover. Conclusion: avoid, particularly in LNG and LPG sea tankers

The thermal overfill energy associated with this type of stratification is negative, hence "thermal underfill".

Thermal underfill is dangerous to have in a storage tank because mixing will lead to sub-atmospheric pressure in the tank and the consequential ingress of atmospheric air, or the collapse of the tank which may not be designed to withstand a negative pressure differential.

Filling an empty tank with thermally subcooled liquid will also lead to auto-stratification. Some of the subcooled liquid will absorb heat from the tank walls, piping, etc. to produce significant quantities of saturated liquid at the tank operating pressure, which partially evaporates as cooldown vapour. This saturated liquid will form a warmer, less dense stratified layer on top of the thermally subcooled liquid.

Filling a partly filled tank with thermally subcooled liquid may result in possibly three layers, with a top saturated layer, a middle layer of heel liquid, and a bottom layer of subcooled liquid.

Thermally subcooled liquid must not be loaded into sea-going LNG or LPG tankers. Otherwise, all will appear to be well when the tanker sets sail. However, when the sea gets rough outside the harbour, the motion of the tanker will induce mixing and generate a negative pressure in the tanks, requiring emergency pressurisation from a tank with sufficient vapour pressure, or using Nitrogen or "Inert Gas" from a ship-board production plant.

6.7   Prevention and Avoidance of Stratification, and Hence Rollover

The prevention and avoidance of density stratification, and hence avoiding rollover during cryogenic storage at 1 bar, may be achieved with the help of the following advisory list:

(1) Correct custody management should enable rollover to be avoided.
(2) Adopting loading, filling and refuelling procedures to produce effective mixing with heel already in the tank, to try to avoid creating stratification in the first place.
(3) If stratification is suspected after loading, adopt 100 % mixing procedures without delay.
(4) Appropriate instrumentation is needed to detect and monitor stratification in each and every tank.
(5) Adequate design and installation is needed of tank auxiliaries, bottom-fill nozzles and top-fill sprays, tank vents, vapour lines, and cross links, to aid rapid mixing before rollover can occur.
(6) Possible use of internal convective and mechanical mixing devices should be considered.
(7) If these steps fail to prevent a rollover taking place, then be prepared to open all vents to the atmosphere as soon as the rising BOR and associated tank pressure exceeds the capacity of the vapour return lines, in anticipation of a Mode 2 rollover with very high BOR.

6.8   Instrumentation for Detection of Stratification

In the absence of precision instruments, a decrease in BOR with time, below the normal boil-off, could be the first sign of stratification in a tank, and possibly be the only indication available towards deciding to commence a mixing operation.

High precision instruments are required to detect density, temperature, and composition changes of the order of 1 %, while location of the liquid-liquid interface requires multi-point measuring heads, or a vertically traversing head, all measuring to 0.1 % precision. This level of precision is a challenge. Measurement of temperature to 0.1 % can be achieved in the laboratory, but considerable effort is required to meet this level of precision under industrial conditions. The precision of normal industrial thermometers is in the range of 1–2 %.

The measurement of density with in-situ instruments to the required precision of 0.1 % is difficult to achieve—the advent of cold electronics should help [4].

Measurement of composition by liquid withdrawal through a capillary tube, followed by total evaporation and gas phase analysis, is extremely time consuming and expensive for any continuous monitoring programme to be maintained. Furthermore the accuracy will be plagued by the Marangoni film flows.

The interface may be detectable using ultrasonic or electromagnetic (optical, radio, etc.) waves. It is therefore possible that a cryogenic sonar/radar system, using low temperature electronics, could be devised to detect stratification.

Density stratification, under constant pressure at 1 bar, is accompanied by a small but progressive increase in liquid volume as the thermal overfill increases in the lower layer. An observed rise in this small liquid level with time is a sure sign of stratification.

6.9   Adequate Provision of Tank Auxiliaries

Tank systems must enable the operations to be carried out for dealing with and removing a build-up of thermal overfill through stratification, before a rollover or vapour explosion occurs. All tanks should therefore have bottom-fill capability via angled nozzle injectors, top-fill capability via spray nozzles sited well above the highest liquid level, and bottom-exit discharge pump lines feeding from a sump below the bottom of the tank, and cross over connections to the spray filling nozzles.

The angle of the bottom-fill nozzle injector must be such as to induce both horizontal and vertical swirl motions to the fresh liquid to achieve satisfactory mixing. The vertical swirl must be strong enough to reach the surface of liquid in the tank right up to the point when the filling operation is complete. For top filling, the spray nozzles must be widely distributed so that the fresh liquid falls evenly across the whole surface of the liquid content.

When a rollover occurs, the vapour lines and emergency vent valves must be large enough for the expected maximum evaporation rate not to raise the tank pressure above the design pressure. A necessary comfort for liquid management!!

6.10   Mixing Procedures to Remove Stratification

6.10.1   Mix as You Fill

Stratification can be avoided by correct and adequate mixing of the tank contents, whether LNG, or LPG, during filling operations. Mixing of fresh, lower density liquid with old weathered liquid can be achieved by bottom filling, provided the filling nozzles are pointing diagonally upwards. Instrumental monitoring is needed to test the effectiveness of filling with mixing procedures; otherwise guessing is totally inadequate and possibly dangerous.

6.10.2   Mixing Will Produce Additional Vapour Flow

Once stratification has occurred, and has been detected, the extra thermal overfill of the lower layer will have to be removed with some significant additional evaporation, the vapour volume generated being possibly path-dependent. This requires careful and safe tank management by mixing, either internally, or preferentially by liquid transfer between two cold tanks. If the mixing is too rapid, the additional evaporated gas flow may exceed the capacity of the vapour return lines, the pressure will rise and the safety vents will open allowing large quantities of flammable gas into the environment. Before this happens, the mixing should be slowed down or stopped until the excess pressure is reduced and brought under control.

6.10.3   No Mixing, to Be Avoided

The practice of adding fresh liquid without any mixing, so as to store two stratified layers with different density at 1 bar in a single container or tank, is dangerous and unnecessary. This practice, in the case of LPG, was advisedly stopped in the 1960s, long before the first LNG rollover events were met in the 1970s.

6.10.4   Action If Stratification Is Suspected

In addition to management of effective mixing, the real problem arises in deciding what to do, if it is suspected that there is still some stratification present in one or more tanks, after filling is completed. If nothing is done, then a rollover will undoubtedly occur, if the suspicion is correct.

On a tank farm with several storage tanks, one effective, and popular, procedure is to transfer liquid intermittently or continuously from tank to tank via bottom emptying and top spray filling.

Passage through the spray nozzles results in some pressure drop and partial evaporation and self cooling of the liquid. The liquid spray droplets have increased density and consequently sink through the top layer liquid so as to promote mixing. The additional vapour generated arises from the release of thermal overfill energy arising from the stratification, together with the heat in-leak and pump energy input during the tank-to-tank transfer.

The tank vapour pressure is determined by the rate of production of this additional vapour and can therefore be controlled, in principle, by the rate of liquid transfer. Should the vapour space pressure rise too high, the liquid pumps should be stopped to try and reduce the rate of dissipation of thermal overfill energy, and allow the vapour pressure to fall back.

On a multi-tank farm, the safest procedure is to carry out tank-to-tank transfers continuously, when the overall rate of vapour generation should become constant. Initially, inter-tank mixing will produce a high rate of evaporation as any thermal overfill introduced during a fill with fresh, dense liquid is dissipated.

With a single tank in service, the only effective option is to circulate liquid out from the bottom and in through the top fill sprays. Again, the tank vapour pressure rise is proportional to the additional evaporation produced by heat in-leak during the liquid transfer and the rate of dissipation of thermal overfill. It can again, in principle, be controlled by the rate of liquid transfer, and the pumps stopped if the pressure rises out of control. However, the circulation must be restarted, perhaps slowly at first, in order to totally remove the thermal overfill energy that has built up. There is no other option!!

If vapour is lost through the vents as a result of recirculating liquid, then this is a relatively small penalty to pay for removing a potential rollover when a much larger quantity of vapour would be lost.

6.10.5   Care When Topping up to 99 % Full

Finally, one word of warning when handling a tank or vessel which is close to full, and being topped up. Remember, that in addition to the intrinsic errors of the high liquid level alarms, that LNG has a significant volume expansion with increasing temperature, and volume compression with increasing pressure, like all cryogenic liquids (see Table 5.1). Under no circumstance must an excess tank pressure, or vapour return line pressure, be controlled by suddenly switching on vapour compressors to absorb an increasing vapour flow.

The vapour line pressure and hence the tank vapour pressure may be inadvertently reduced so as to cause liquid to expand and boil up into the vapour lines. The subsequent uncontrolled increase in evaporation rate and rise in pressure may lead to serious mechanical damage. Switching on vapour compressors should be carried out very gently when the vessel is brimful, and then only intermittently to start with.

6.11   Path Dependent Mixing of Boiling Cryogenic Liquids

So far in this chapter, we have not considered the volumes of vapour generated when two boiling liquids are mixed together. Let us now look more closely at the mixing process, particularly when the two liquids have widely separated boiling points.

At temperatures above ambient, the homogeneous mixing of boiling liquids is not an everyday experience. Adding immiscible liquids, like boiling oil into water, and vice-versa, are probably the closest we meet, when the results are spectacularly explosive and hazardous because of homogeneous nucleate boiling of the water.

On the other hand, at temperatures below ambient, the forced convection mixing of miscible boiling cryogenic liquids, such as LNGs, LPGs and other hydrocarbon liquids, is commonly carried out. Intuitively, one might expect the vapour generated to contain more of the lower boiling point component, but hardly that the volume of vapour might vary significantly with the mixing profile.

The mixing process is, of course, irreversible, and is accompanied by an irreversible increase in entropy. Mixing with evaporation is particularly difficult to model because, in addition to the thermal mixing and release of thermal overfill, there is the compositional heat of mixing with irreversible entropy production which is significantly path dependent. A considerable volume of vapour is produced by the usual positive heat of compositional mixing and by thermal contact between colder and hotter components (no homogeneous nucleate boiling has been observed) before the final equilibrium state of the mixture is achieved.

However, because the mixing is path-dependent, the volume of vapour produced is a variable path-dependent phenomenon. This path dependence is clearly demonstrated when liquid propane and liquid butane are mixed and the vapour produced has to be recondensed by limited available refrigeration. Briefly, when colder liquid (propane rich) is added to hotter liquid (butane rich), the vapour generated is found experimentally to be twice that produced by the alternative when hot is added to cold [5].

It follows that path-dependent mixing of LNG and other cryogenic liquids can be expected to produce significantly large variations in the volume of vapour generated. For example, during the spray mixing of cold LNG into hotter liquid, the vapour generated will be more than if the hotter LNG is spray-mixed into the colder liquid.

6.12   Some Consequences of Path-Dependent Mixing

The mixing experiments at Southampton demonstrated conclusively that vapour flash volumes are path-dependent, and may be expected to be much larger than those predicted from heats of mixing. It follows that all cryogenic liquid mixing can be expected to be path-dependent, with noticeable consequences on operations at large scales, including the following examples:

1. Adding and mixing fresh liquid to the existing heel, to prevent stratification. If cold is added to hotter liquid, the vapour flash volume may be greater than if hot is added to cold liquid. This applies to any pair of cryogenic liquids being mixed, whether LNG or LPG.
2. When removing stratification by mixing the two layers. Bottom to top mixing may generate more vapour flash than top to bottom mixing.
3. When a rollover occurs, the mixing is irreversible and therefore path-dependent. The spontaneous penetrative mixing is from bottom to top, and the vapour flash volume can therefore be expected to be larger than that predicted from the heat of mixing and thermal overfill of the two layers.
4. Since the heat of mixing contains a thermal equilibration between two layers together with an irreversible increase in entropy due to compositional equilibration, computer modelling of rollover BORs can never be very accurate. Every rollover produced at Southampton was different and could only be correlated between Mode 1 events (slow rise by about 5–10 fold increase of BOR) and Mode 2 events (fast rise by between 10 and 100+ fold increase in BOR).

6.13   Possible Use of Internal Mixing Devices to Destabilise Stratification

With a density difference of the order of 1 % between stratified layers, it is possible to envisage the use of convective devices to reduce the rate of build-up of thermal overfill in the lower layer and to encourage convective mixing.

Bearing in mind that once stratification occurs, the 'A' heat in-flow to the lower layer through the tank wall and floor insulation is trapped within the layer. The purpose of a convective device would be to funnel this 'A' heat in-flow up to the top layer, thereby reducing the build-up of thermal overfill in the lower layer and delaying the onset of rollover. The question then is whether the device is effective and practicable.

In tanks with a depth/diameter ratio greater than one, the wall boundary layer suction is strong enough to pull the heated liquid in contact with the tank floor into the wall flow. Experimental attempts to increase the momentum of natural convective wall boundary layer flows in an open vessel have not been successful in the past. It is therefore unlikely that any convective device would be successful in such tanks.

In tanks with a depth/diameter ratio less than 0.5 (as in very large LNG and LPG tanks), the wall boundary layer flow may not be strong enough, and some of the heat inflow from the tank floor may be concentrated by natural convection into a number of thermals rising through the lower layer. As mentioned in Sects. 3.5.3 and 4.5, these thermals are the centres of large convection cells with horizontal spacings of the same order as their vertical dimension, namely the depth of the lower layer. Their number and position relative to the tank floor will therefore also depend on the depth of the lower layer.

If these convection thermals can be caught in large conical apertures leading into chimneys extending through the lower/upper layer interface up to the free liquid surface, then in principle the thermal catcher and chimney device could be effective in reducing thermal overfill in the lower layer [6]. The devices would have to be positioned around the tank floor and not in the centre where the central downward jet from the upper layer surface would act in opposition.

Another device includes the possible multiple concentration of the heat in-flow through the tank floor to provide hot-spots for generating hot thermals with high buoyancy capable of passing through the lower/upper layer interface to the free surface without the aid of convective chimneys.

Devices using the introduction of gas-bubble streams are unlikely to be efficient mixers in large tanks with liquid depths of 10–50 m.

All these internal convective mixing devices may work well on a small scale in the laboratory, but no reports have been seen of their use in large scale storage tanks.

The obvious alternative for internal mixing is the use of power-driven rotating paddles in each tank. If submerged electric motors could not be approved as safe for use, then motors driven by, say, pressurised nitrogen gas could be considered.

In the LNG safety laboratory, Southampton University, magnetic stirrers have been in common use for all mixing operations, using magnets rotating outside the LNG container.

6.14   Some Comments on Handling an LNG Rollover

1. First of all, read Sect. 5.3 on the fast sequence of uncontrolled happenings in a rollover event, so as to anticipate actions needed, as the sequence develops. Do not panic!!
2. If the pressure and BOR in a tank starts to rise unexpectedly, then all the valves to the dedicated vapour lines should be opened. This should be carried out as soon as possible in anticipation that the BOR will rise 5–10 fold, following the Mode 1 sequence.
3. If the pressure continues to rise, and the vapour lines cannot accept the increasing BOR, then all the vents to the atmosphere should be opened early, in anticipation of Mode 1 converting rapidly (in a few seconds) to a Mode 2 rollover with a much higher BOR.
4. There are no working correlations to help determine whether a conversion from Mode 1 to Mode 2 rollover will happen. Therefore, it is advisable to anticipate the conversion will take place, by having all the available vents open to the atmosphere, when the rising BOR exceeds the vapour line limits.

5. The size of the vents should be large enough to be able to exhaust a 200 fold increase in BOR without approaching the choked (sonic) flow condition. If not, then a Mode 2 rollover could lead to structural failure.
6. With vents open to the atmosphere, a cold dense white fog cloud will form down-wind along the ground or sea for several hundred meters. This visible cloud is actually water vapour condensed from the moist atmosphere by the cold LNG vapour. LNG vapour is essentially invisible but as a source of cold it remains within the confines of this visible cloud. As the cloud continues to move downwind, it mixes and warms up with the air. It will eventually become less dense than atmospheric air and rise rapidly away from ground level, or sea surface, and clear of potential ignition sources. If the entire fog cloud is treated as potentially flammable, this assumption will be on the safe side.
7. If the rollover is on a tanker, the ship should be turned across the wind so that the vented vapour cloud is carried away from on-board ignition sources.
8. With vents fully open to the atmosphere, there is now time to consider second-ary effects of the rollover on neighbouring tanks, such as varying pressures in common vapour return lines.
9. Some isolation, from other similar tanks nearby, or on the same tanker and with the same recent history of custody management, needs to be considered to prevent triggering further rollovers.
10. In addition, if there are tanks which are full to the brim, they must not be exposed to rapid changes in vapour line pressures, so that liquid enters the vapour line.
11. Eventually, the BOR will start to fall steadily, but this final part of the sequence may take several hours before the BOR decreases back to the normal unstratified level.

6.15   Summary of Properties of LNG and LPG Mixtures, Relating to Stratification and Rollover

1. (T–x) data for surface evaporation of LNG and LPG mixtures may not be the same as the (T–x) free-boiling data used for distillation. The deviation in vapour composition from the equilibrium data is proportional to the surface evaporation mass flux.
2. While stratification in a single component liquid may lead to QHN boiling or vapour explosion, the consequences of density stratification at 1 bar are more serious in multi-component liquid mixtures. The density varies with both temperature and composition, and stratified layers experience double diffusive convection instabilities and spontaneous mixing or rollover.
3. The inevitable consequence of stratification in a mixture at 1 bar is a rollover incident when the two layers mix spontaneously by vertical oscillating penetra-tive convection plumes across the whole of the liquid/liquid interface.

4. The peak boil-off during rollover is determined by:

   (a) the increased superheat of the surface layer in Mode 1 rollover, or
   (b) the vertically oscillating, penetrative convection reaching the surface sub-layer in Mode 2 rollover, and breaking down it's morphology so that the full unimpeded mass flux of molecular evaporation is reached at 20–250 times the normal boil-off, depending on the magnitude of the evaporation coefficient.

5. There are several ways in which autostratification can occur.
6. Monitoring of every tank for the occurrence of stratification is necessary, otherwise continuous transfer of the liquid between tanks should be applied to promote mixing as a preventative measure.
7. Once stratification takes place, efficient mixing should be promoted to stop rollover occurring.
8. If the sequence of increasing BOR and rising tank pressure starts to take place for no apparent reason, then the effort to get rid of stratification has failed, and an uncontrolled rollover event should be anticipated. All vapour valves and vents should be ready for immediate opening to minimize pressure rise as the BOR rises and rises.
9. This summary of properties applies to all cryogenic liquid mixtures, including LNGs, LPGs or other hydrocarbon mixtures.

References

1. SIGTTO: Guidance for the Prevention of Rollover in LNG Ships. Witherby, Edinburgh (2013)
2. Booth, D.A., Bulsara, A., Joyce, F.G., Morton, I.P., Scurlock, R.G.: Wall film flow effects with LNG. Cryogenics **14**, 562 (1974)
3. San Roman, O.: The dynamics of methane/ethane separation by differential surface tension driven flows; or the Marangoni effect. PhD thesis, Southampton University (1978)
4. Scurlock, R.G.: On-line instrumentation of cryogenic systems and plant to an accuracy of 0.01 % using cold electronics. In: Proc. ISA, Houston, vol. 32, p. 139 (1993)
5. Tchikou, A.: Mixing of propane and butane. PhD thesis, Southampton University (1985)
6. Voyteshonok, V.: Safe storage vessels for low-boiling liquids. Cold Facts **21**, 33 (2005)

Chapter 7
Vacuum Insulated Tanks for Pressurised LNG

Abstract The rise of LNG as a green fuel, in the face of global warming, is leading to the use of relatively small, vacuum insulated VI tanks for 1–100 m³ capacity of pressurised LNG. Such tanks have been widely used over many years for the supply of nitrogen, oxygen and argon as cryogenic liquids. They are called 'zero-boil-off Liquid Gas Cylinders', and they have never been a problem for unstable evaporation of the cryogenic liquid contents. This chapter describes how this handling behaviour also applies to LNG.

7.1 Summary

1. No rollover apparent under storage and handling in zero boil-off, pressurised, VI tanks.
2. Wide experience with handling other cryogenic liquids under pressurised, zero loss, storage, such as LIN, LOX, and LAR.
3. Stable stratification with zero loss under pressure, and no rollover.
4. 100 % mixing advised during refuelling and refilling.
5. Density reduction and liquid expansion with increasing temperature and pressure require only 85 % filling.
6. Possible problems.

7.2 Introduction

There appears to be NO circumstances under which LNG under pressure from zero-loss, pressurised storage in Vacuum Insulated (VI) tanks can suffer Rollover. The only provisions are that during refilling and refuelling operations, complete mixing of fresh LNG with the tank heel should be achieved, before the refilling to only 85 % full is completed.

© The Author(s) 2016
R.G. Scurlock, *Stratification, Rollover and Handling of LNG,*
LPG and Other Cryogenic Liquid Mixtures, SpringerBriefs in Energy,
DOI 10.1007/978-3-319-20696-7_7

In practice, the mixing provision may not be necessary for mobile tanks, because sloshing of the tank contents, arising from accelerated motions of the road vehicle or ship, will produce the required mixing.

7.3   VI Tanks Widely Used for Cryogenic Liquids

Double-walled VI tanks have been widely used for many years on an industrial scale for the pressurised storage, transport and controlled supply of liquefied gases such as LOX, LIN, LA. Originally perlite or other expanded powder were used as the filler and insulation between the double walls, but the more recent development of multi-layer reflective insulation, MLI, together with reliable high vacuum, has enabled much lower heat influxes and boil-off rates to be obtained. The extensive experience with pressurised, standard cryogens makes VI tanks particularly ideal for storage and handling of LNG.

7.4   Zero Boil-off Loss Storage Under Pressure

One disadvantage of double-walled VI tanks is that the outer case has to withstand the collapsing pressure of the atmosphere at 1 bar and is therefore rather heavy. On the other hand, the inner liquid container is a pressure vessel which can easily have a working pressure up to 6–25 bar in 100–500 L volume truck fuel tanks; working pressures of 10–15 bar in tanks of 1–100 m^3 volume suitable for small ship bunker fuel tanks; and working pressures of 5–8 bar in larger tanks of 100–1000 m^3 for ship and shore fuel stores. This pressure vessel capability allows zero-loss storage, as the pressure slowly rises to the safety or working limit over a period of 50–100 days.

7.5   Stable Stratification and No Rollover

Under the isobaric (constant pressure) operation of a 1 bar tank, the heat flow into the liquid is removed from the tank continuously by the latent heat of vaporisation of the boil-off vapour, apart from the portion entering the lower layer when there is two level stratification.

In a zero loss VI tank, the dynamics of the heat flows into the liquid contents are quite different, being isochoric (constant volume) in character. In static tanks with no sloshing, the total heat inflow is absorbed by the stratified upper layer of the liquid. Experimental studies by Chrz and Suma [1], have shown that the stratification is extremely stable whereby the upper layer of warm saturated liquid at T2, with a saturation pressure at P2, grows hotter with time, above a colder layer at the initial (filling or refuelling time) saturation temperature T1. The heat flow through the tank walls is carried by boundary layer flows into the upper layer causing T2 and P2 to increase with time. With sloshing of the contents tending to mix the upper and lower layers, the rate of increase of T2 and P2 with time is less.

Experimental studies using LIN in three static VI tanks, of volume 6, 11 and 31 m³ respectively, demonstrated, that the rate of increase of T2 and P2 with time was just over two times the rate expected for uniform heating of the whole tank contents with no stratification (Fig. 7.1).

In other words, and more specifically, the mass of the upper liquid layer, which was absorbing all the heat inflows, remained at an average of 44 % (between limits of 38 and 52 %) of the tank volume, during measured and calculated hold times with zero-loss storage, while the lower 56 % liquid layer remained at the initial saturation temperature of T1.

While the temperature difference (T2 − T1) between the layers is quite large, the isochoric density difference between the two layers is relatively small, with no trapping of heated boundary layer flow in the lower layer. The stratification is therefore convectively stable, with no build-up of thermal overfill in the lower layer, and there appears to be no chance of double-diffusive convection creating a spontaneous mixing and rollover.

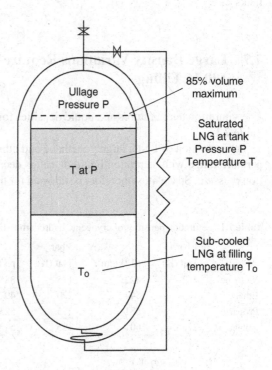

Fig. 7.1 Two layer density stratification in LNG under pressure within VI tank. *Upper layer* 44 % average, at saturation temperature T and ullage space vapour pressure P. *Lower layer* 56 % average, remaining at filling temperature T₀ but sub-cooled under ullage pressure P+hydraulic head

Ullage Pressure P

T at P

T₀

85% volume maximum

Saturated LNG at tank Pressure P Temperature T

Sub-cooled LNG at filling temperature T₀

7.6   100 % Mixing Requirement for Static Tanks

With single component cryogenic liquids LIN, LOX, LA, etc. in VI tanks, under isochoric conditions, there is no storage instability problem from refilling with colder liquid, which will settle below the warmer heel. If mixing is not complete, the stratification of warm layer above cold layer will take place as normal during isochoric, zero-loss storage, when the tank contents pressurise as usual within the tank volume.

However, with LNGs having different compositions as well as temperatures, it appears advisable that 100 % mixing practices are adopted during refuelling and refills, to prevent the initial formation of multiple layers of liquid. When the LNG is subsequently pressurised, with zero loss of vapour, there is then no chance of double-diffusive convection leading to spontaneous mixing and rollover between adjacent layers.

This mixing practice appears to be a requirement for trouble-free storage of LNG under pressure. Complete 100 % mixing will, of course, ensure that normal hot–above–cold, two layer stratification will take place during pressurisation under stable convective conditions.

It is, however, well known that (a) sloshing of the tank contents is generated by accelerated motions of the road truck or ship and (b) sloshing will aid the mixing process.

So this advisory provision on mixing practice may therefore only apply to static tanks.

7.7   Large Density Variations Require Less than 100 % Filling

The saturation temperatures and liquid densities for different pressures are shown in Table 7.1.

Pressurisation of LNG, largely methane and ethane, takes place over a wide temperature range with saturation liquid densities decreasing despite increasing saturation pressure. Space must therefore be allowed for the expansion of the upper layer of

Table 7.1 Saturation densities of cryogenic hydrocarbon liquids at 1, 5 and 10 bars

	1 bar		5 bar		10 bar	
	Tsat (K)	$\rho 1$ (kg/m³)	Tsat (K)	$\rho 5$ (kg/m³)	Tsat (K)	$\rho 10$ (kg/m³)
Methane	112	422	135	385	149	359
Ethane	185	544	220	497	241	464
Propane	231	581	275	525	300	489
Butane	273	601	323	543	353	501

stratified LNG as the pressure builds, as well as contraction of the vapour space above the liquid. Under no circumstances should the vapour space reduce to zero as the liquid expands with increasing saturation pressure. Adequate vapour volume must exist so that, if the vent/safety valve lifts, there is no venting of liquid from the tank.

The maximum allowed filling is a function of the initial saturation pressure of the liquid as delivered into the tank. For safety reasons, the delivered liquid is considered to be at 1 bar absolute, and the maximum saturation at the relief valve set pressure, independent of the intended storage time. From the table, the maximum filling is 85 % to allow expansion to 100 % tank volume, as allowed by US standard NFPA 59A. Other standards require some remaining ullage volume after full expansion, e.g. an additional 5 %. The experiments by Vaclav and Suma indicate that, for static tanks, not more than 50 % of the tank liquid suffers expansion to saturation at the relief valve pressure. It therefore follows that 85 % filling is a safe figure for static tanks, exceeding the actual expansion by a factor of around two.

7.8   The Stratification Process During Pressurisation and Zero Loss

Under isochoric (constant volume) conditions for the whole tank contents, about 98–99 % of the heat-flow energy, entering the liquid through the tank-wall insulation, appears to be absorbed by increasing the enthalpy of the upper layer of liquid. A small amount, of the order of 1–2 % of the heat flow, is absorbed by the latent heat of vaporisation of a small amount of liquid to provide the vapour needed to increase the pressure.

Thereby, there is an equilibrium between the rising saturation pressure in the ullage (vapour) space, together with the rising surface temperature of the evaporating liquid. As a consequence, the heat energy is retained within the tank as the pressure rises slowly towards the specified maximum pressure (~8–10–24 bar), determined by the relief valve setting.

Looking more closely at stratification in a VI tank, we note that "A1" heat fluxes entering the upper layer of liquid are carried by convection to the liquid/vapour surface by superheated boundary layer flows up the walls of the tank. A small part of the energy (1–2 %) is absorbed by evaporation, while the bulk of the energy remains in the upper liquid layer of superheated (with respect to T1) liquid, forming a stable upper layer at a higher temperature T2.

The "B" heat fluxes into the unwetted upper section of the tank enter the ullage vapour helping to raise its temperature and pressure.

The effects of the "B" heat flows in raising the tank pressure can also be enhanced by an external loop, between bottom and top of the tank, of liquid evaporating in an external heat exchanger via a pressure raising control valve. When a flow from the storage tank of gas or liquid is demanded, the pressure raising valve and heat exchanger in the loop can be controlled automatically.

The "A2" heat flux entering the lower liquid layer is carried by convection to the upper/lower liquid layer interface by boundary layer flows at the tank wall. Because the density difference between the layers remains small under the isochoric conditions, the buoyancy of this boundary layer flow continues to be sufficient to penetrate into the upper layer and mix with it. In this way, the lower layer is not heated by the insulation heat inflows and remains essentially at the refuelling temperature T1 of the original mixture at 1 bar.

This stratification process is convectively stable, and there appears to be no possibility of spontaneous convective mixing, i.e., there is NO rollover in a static VI tank.

The stable stratification may also be used as a stable dynamic property to control the rising gas pressure, as well as for controlling the feed supply of gas or liquid out of the tank.

This property of VI tanks is widely used in cryogenic engineering applications using LIN, LOX and LA. It is applicable in a similar way for using LNG as a liquefied fuel for road trucks and ships at pressures up to 8, 10, 15 or 24 bar.

7.9 Possible Problems

7.9.1 Marangoni Films

It may not be realised that the unwetted walls above the stratified LNG will be covered with differential surface-tension-driven Marangoni films of ethane rich liquid up to the saturation temperature of ethane, and propane rich liquid up to the saturation temperature of propane. Ethane and propane rich droplets will run down the walls against the film flow and collect in the bottom of the tank, as an example of auto-stratification of a high density layer at the bottom of the tank.

Cold instrumentation heads inside the tank will give incorrect readings if they are covered with Marangoni film.

7.9.2 Effect of Sub-cooling Difference Between Layers

With large temperature differences between the upper and lower layers, there will be substantial reduction in sub-cooling of liquid taken from the bottom of the tank, as the liquid level drops. When the exit flow of cold lower layer is replaced by the flow of warmer upper layer, the reduction in thermal sub-cooling may cause a liquid pump to stall, or a long liquid line to stop flowing, through two-phase flow inducing excess pressure drops within the pump or along the line. Raising the tank pressure should enable the exit flow to be renewed.

This problem has been widely experienced, when any stratified liquid is taken from a pressurised VI tank. One quick answer is to raise the tank pressure by feeding back from the pump exit into the vapour space.

7.9.3 Beware of Pressure Cycling Non-condensing Purge Gas to Empty a Tank

The pressure cycling of a non-condensing purge gas, such as nitrogen or helium, to completely empty a tank of residual LNG for maintenance purposes, should be approached with caution.

The non-condensing gas will tend to reduce the partial-pressure of residual natural gas vapour below the saturation pressure of any residual LNG liquid left in the tank. In addition, the purge gas may dissolve in the surface layers of the residual liquid, creating an impedance and restriction to the required surface evaporation.

The liquid is then superheated, with respect to its partial pressure and temperature, and can boil explosively as a QHN event. Such an event may cause mechanical damage to the tank and fittings.

So, avoid non-condensing purge gas to pressure cycle and thereby empty LNG fuel tanks: instead, use natural gas for purging liquid volumes, and switch to nitrogen only after all liquid has been removed.

Reference

1. Chrz, V., Suma, J.: Dynamics of tank pressure during storage of cryogenic liquids. In: Proc. IIR/ICR 2007, Beijing (2007)

Chapter 8
Liquid Transfers Avoiding 2-Phase Flow

Abstract The transfer of cryogenic liquids like LNG down pipes is not as simple as pumping water. The difference is that cryogenic liquids are stored at their boiling points, whilst water is stored at ambient temperature, which is a long way from its boiling point at 100 °C.

Pumping the boiling liquid can easily lead to much vapour generation, and 2-phase flow. The result is that the pumped mass flow reduces to a minimum or to zero.

The simple way to stop this transfer disaster is to use pressure sub-cooling of the liquid.

8.1 Summary

1. 2-Phase flow explained as a mixture of saturated vapour and liquid, with greatly reduced mass flow.
2. How 2-phase flow can occur.
3. Prevention with adequate sub-cooling by pressure application to counter NPSH of pump.
4. Liquid transfer with transient 2-phase flow—cooldown of long pipeline.
5. Avoiding pressure surges caused by rapid opening/closing of liquid valves.
6. Care with topping up tanks when 99 % full.
7. Zero delivery.

8.2 General Remarks on Subcooled Liquids and 2-Phase Flow

We are all used to handling and transferring water, whether we are using a hosepipe to wash the car, or filling a kettle, or turning taps to have a bath or shower. In all cases, whether the water comes from the cold tap at ambient temperature or from the hot tap at, say, 60 °C, the water is subcooled way below its normal boiling point

R.G. Scurlock, *Stratification, Rollover and Handling of LNG,*
LPG and Other Cryogenic Liquid Mixtures, SpringerBriefs in Energy,
DOI 10.1007/978-3-319-20696-7_8

at 100 °C. It is subject to a pressure, whether it is atmospheric pressure or a greater hydrostatic pressure, which is well above its saturation vapour pressure. The water transfers are never a problem because there is no 2-phase flow — unless dissolved air comes out of solution, when there may be a hint of problems like water-hammer, irregular and reduced flow, and so on.

Handling and transferring a volatile liquid, such as petrol, is not quite so easy because the normal degree of subcooling is not so large as with water. Storing petrol at sub-ambient temperature in underground tanks at filling stations ensures there are no 2-phase problems when filling a road vehicle's fuel tank. However, a vapour lock in a hot engine cutting out the fuel supply is a good example of the sub-cooling being lost in the fuel supply line, thereby creating a 2-phase problem.

Transferring LNG and LPG is straightforward as long as the liquids are above saturation pressure at the transfer device (e.g. a pump right through to the end of any transfer process). However when 2-phase flow occurs, the problem is a major one and the transfer may slow down, or stop altogether.

8.3   What Is 2-Phase Flow?

When any liquid evaporates, whether cryogenic or otherwise, a very large volume of vapour is generated at the same pressure. In the case of liquid methane or LNG, the volume of vapour at ambient temperature and pressure is some 600 times the volume of the liquid before it evaporates. This means that, say, if only 1 % of the mass of the liquid in a transfer line evaporates, the volume occupied by the vapour is six times the total volume of unevaporated liquid. Although the vapour mass is 99 times less than the liquid mass, the vapour volume occupies six sevenths of the volume of the line, and the mass flow almost stops.

In general, a mixture of vapour and liquid, having a much lower density, must have a much greater velocity in order to maintain a required mass flow rate. The limiting velocity is, of course, the local velocity of sound in the fluid.

While the velocity of sound in cryogenic vapours is quite high ($\sim$200–300 m/s and proportional to $T^{0.5}$), it is much lower in any 2-phase mixture because of the high adiabatic compressibilities of all mixtures.

Useful empirical correlations have been developed by Martinelli and Lockhart [1] for calculating frictional pressure drops at ambient temperature, which have been extended with reasonable accuracy to low temperature flows [2, 3]. The net result is that the occurrence of 2-phase flow with a fixed available overpressure for liquid transfer, will lead to a significant fall in mass flow and a mass transfer rate close to zero.

Likewise, if 2-phase flow develops in a rotary liquid transfer pump, the mass flow output will be reduced or the pump may fail to prime altogether, leading to overheating and mechanical failure. Transient or continuous oscillations may also lead to mechanical damage.

The obvious step is to avoid 2-phase flow but, first, there is a need to understand how 2-phase flow happens.

8.4   Occurrence of 2-Phase Flow

Let us consider the P–T diagram in Fig. 8.1, where XY is the saturation vapour pressure-temperature line, or alternatively the pressure versus boiling point curve. In addition to showing the saturation vapour pressure curve, the diagram can also be regarded as a thermodynamic state diagram, with the curve separating liquid and vapour phase thermodynamic states.

- All points above the curve represent liquid phase states,
- all points below represent vapour phase states,
- and all points on the curve represent liquid and vapour in contact with one another, namely the 2-phase state.

Consider liquid with thermodynamic state 'A' at pressure P and temperature T. The liquid is undercooled, or subcooled, with respect to its boiling point at pressure P.

During a transfer operation, the liquid state can cross the saturation vapour pressure curve XY resulting in the creation of two phases by two separable paths, or by combinations of the two paths, i.e.,

(1) by a reduction in pressure, or pressure subcooling, along AB (or strictly along AB'),
(2) by absorption of heat, producing a rise in temperature at constant pressure, along AC, i.e. a reduction in thermal subcooling,
(3) by various combinations of pressure reduction and heat absorption.

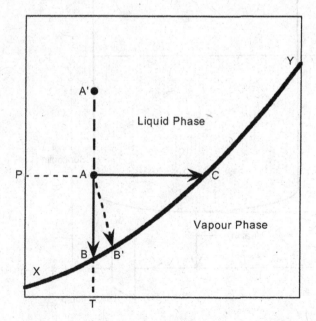

Fig. 8.1 P–T diagram showing saturation vapour pressure versus temperature curve, separating liquid and vapour thermodynamic states

To prevent the occurrence of 2-phase flow during a liquid transfer, the change in thermodynamic states represented by both paths AB and AC must not end on, or cross, the saturation vapour pressure curve XY.

This can be simply achieved with adequate pressure subcooling of the liquid, by pressurising the liquid along AA′ before the transfer commences.

8.5 Pumped Liquid Transfer Avoiding 2-Phase Flow

Consider Fig. 8.2 where we have a storage tank containing liquid, density ρ, depth L, stored at pressure Ps and saturation temperature Ts with a pump mounted a vertical distance H below the bottom of the tank. Liquid entering the pump is subcooled by the hydrostatic pressure head $\rho g \, (H+L)$, but loses some pressure through frictional pressure drop and gains heat through the wall of the connecting line. The changes in thermodynamic state are shown in Fig. 8.3.

Like any liquid pump, whether rotating or reciprocating, the Net Positive Suction Pressure Head (NPSH) is an additional requirement. Clearly, the reduction in liquid pressure as it is sucked into and accelerates through the inlet passages of the pump,

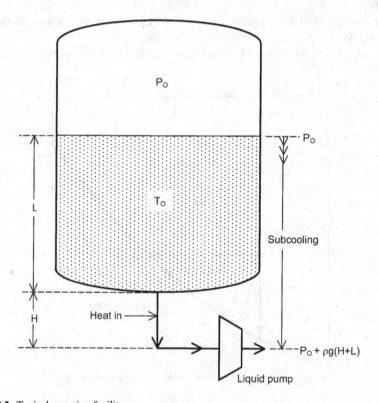

Fig. 8.2 Typical pumping facility

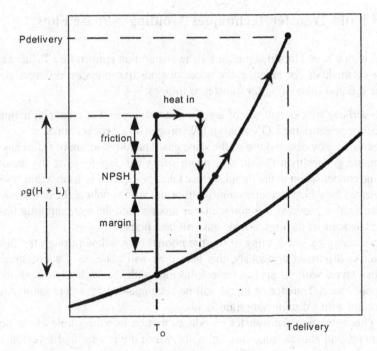

Fig. 8.3 P–T diagram showing change of state of liquid during a pumping operation

namely the NPSH, must not cause the liquid state to cross the saturation line into the
2-phase state. If it does, then cavitation or partial evaporation will occur causing the
pump to stall. The liquid pressure before entry into the pump must therefore always
be greater than the saturation pressure Ps by the magnitude of the NPSH of the pump.

As the pumping process continues, the liquid level falls reducing the hydrostatic
pressure head and associated pressure subcooling, at the entry to the pump. Eventually,
the total pressure head falls to the NPSH or below; the pump begins to stall and will
eventually no longer remove further liquid from the tank. (However, if the outlet
valve of a centrifugal pump is throttled when the liquid level falls to near this stalling
point, the flow reduction in effect reduces the NPSH and enables the liquid to be
pumped further. This can be important if the tank is to be emptied completely.)

This is not the whole story, because, as the liquid level drops, the vapour in the
vapour space expands into the volume previously occupied by the liquid removed.
Generally, there is not enough liquid evaporation to maintain the vapour pressure,
and the pressure falls, causing the pressure subcooling at the pump entry to further
reduce. This latter effect can be alleviated by back-filling vapour into the tank.

For tanks with insulation provided by a gas purged foam or powder at 1 bar, it is
desperately important that a negative pressure is not generated inside the tank by the
liquid pumping process. Back-filling with vapour flashed from liquid in the pump
delivery line should be employed. Otherwise, the negative pressure could lead to
the tank-wall collapsing inwards.

8.6   Liquid Transfer Techniques Avoiding 2-Phase Flow

Starting with a stored liquid at pressure P_0 and saturation temperature T_0 there are a number of methods for creating the necessary positive pressure differential and associated liquid subcooling for transfer as follows:

(1) Pressurising by external use of a non-condensing gas. For example, helium is used for pressurising LOX and liquid hydrogen space rocket tanks.
(2) Pressurising by external use of the same gas. The introduction of warm gas at a pressure greater than P_0 into the vapour space will rapidly raise the pressure. Some condensation at the liquid surface takes place so as to form a thin layer of saturated liquid in thermodynamic equilibrium with vapour at the new pressure. Stratification prevents this surface layer mixing with the rest, and little further condensation of the pressurising gas will take place.
(3) Pressurising by self-heating, using the normal heat inflow through the insulation. As discussed previously, this heat flow will generate a warm stratified upper layer, with its surface in equilibrium with the higher pressure in the vapour space. Transfer of liquid will be accompanied by a continuous fall in pressure with a diminishing transfer rate.

   This is because the transfer introduces motion into the whole of the liquid contents and causes some mixing of the warm upper stratified layer into the colder lower layers. The surface temperature falls, there is some condensation of vapour and the vapour pressure also falls. At the same time, expansion of vapour into the volume previously occupied by the exiting liquid also contributes to the fall in pressure.

   Thus, self-heating is adequate for transferring relatively small quantities of liquid from a large efficiently insulated tank, say 5–10 % in a period of 24 h.
(4) Pressurising by use of a pressure raising coil or vaporiser connected between bottom and top of the vacuum-insulated tank. The flow is driven by the difference in hydrostatic pressure heads between liquid in the tank and vapour in the line above the vaporiser, and may be simply controlled by a valve in the vapour line to the top of the tank. Automatic control via a pressure regulator enables the tank ullage pressure to be maintained independently of the rate of liquid removal.
(5) The use of single stage and multi-stage mechanical pumps, generally rotary, for liquid transfer purposes. These can be:

   (a) submerged in the liquid, are permanently cold and therefore ready for immediate transfer, e.g. for LNG and LPG, or
   (b) externally mounted, and require cooling and priming before transfer can take place.

In all cases, the liquid transfer will stop, or perhaps not even start, if the liquid state crosses the saturation curve and creating 2-phase flow at any point in the transfer process.

   The consequences of a liquid transfer failure can vary from just an inconvenience, resulting in wasted time and cryogenic liquid, to a major fire and explosion in a mechanical pump which overheated when it failed to prime. It is therefore important for operators of liquid transfer equipment to understand the importance of avoiding 2-phase flow by working with an adequate margin of subcooling.

8.7 Liquid Transfer with Transient 2-Phase Flow

During the cooldown of liquid transfer lines and storage tanks or containers, the occurrence of 2-phase flow, albeit transient, is unavoidable.

Now, in steady state 2-phase flow, there are a variety of ways in which vapour and liquid can co-exist in a pipeline, such as slug, mist and annular flows, all of which are well documented in texts on boiling heat transfer and cryo-engineering (see also, Sect. 8.3).

During a cooldown process, there are three flow sections:

(a) liquid in the cold portion of the transfer line
(b) 2-phase flow and cold vapour flow where cooldown is taking place
(c) warm vapour exiting through the vapour lines.

The pressure drop in (a) the length of line filled with cold liquid is relatively small and can generally be ignored.

The pressure drop in (b) the cooldown area is higher but need not be excessive if the flow is not being forced. The heat transfer process which is achieving the cooldown is a mixture of natural and forced convection heat transfer between cryogen and solid surfaces of the cryogenic system and is difficult to model accurately. The main finding is that the heat transfer is only weakly dependent on cryogen velocity and hence pressure drop. Forcing the flow with a high pressure input will not noticeably increase the rate of cooldown.

The pressure drop in (c), the vapour venting section, is by far the largest quantity and is limited by Fanno flow, when the velocity becomes sonic at the exit. This is because the volume flow of warm cryogen vapour is much greater than the volume flow in the 2-phase cooldown section, and is several hundred times greater than the liquid input volume flow.

The best way of understanding cooldown is to consider some examples, namely a long transfer line, a tank and a large mass.

8.8 Cooldown of a Long Pipeline with L/D Greater than 2000

With a long pipeline, length L, inner diameter D, the question is how to cool the pipe down to a liquid temperature and commence transfer, the operation to be as fast and as thermodynamically efficient, as possible.

Precooling a transfer line with a second liquid to an intermediate temperature is not an effective option even if the liquid is conveniently available. It is time consuming and it is also difficult to remove all traces of the second liquid.

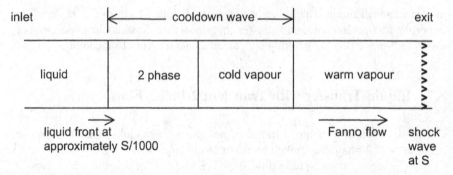

Fig. 8.4 Cooldown of pipeline showing liquid front, cooldown wave, Fanno flow of warm vapour, and shock wave at the exit

The simplest and most effective method is to use the liquid cryogen to be transferred. In addition to the latent heat of evaporation, there is the all-important sensible heat of the vapour from boiling point up to ambient temperature available for cooling the pipeline. As mentioned several times already, this sensible heat is the same magnitude as the latent heat of evaporation for liquid natural gas, and for liquid propane.

Using subcooled liquid from one end, at an entry pressure above saturation pressure, a "cooldown wave" progresses along the pipeline as indicated in Fig. 8.4.

Ahead of the liquid front, 2-phase flow occurs as the inner wall of the pipeline, inner diameter D, is cooled down to the liquid temperature; then there is single phase cold vapour cooling the pipeline to a point about 1000D further down, where the temperature difference between wall and vapour has decreased to zero. The region from this "warm" point back to the liquid front is the so-called "cooldown wave".

Beyond this warm point, the wall and vapour are both at approximately ambient temperature.

The controlling feature of the entire cooldown process is the behaviour of the ambient temperature vapour, as it accelerates down the uncooled portion of pipeline until it reaches the local speed of sound S near the exit. A shockwave develops and all the remaining pressure drop along the pipeline is consumed by increasing the intensity of the shock wave and not by increasing the mass flow.

This flow, or Fanno flow, is well-known to blow-down wind-tunnel operators, in which the maximum mass flow is determined only by the local speed of sound S at ambient temperature, the exit pressure and the cross-sectional area of the exit (see, e.g. [4]).

The velocity of the liquid front behind the cooldown wave is then determined entirely by the Fanno flow of the ambient temperature vapour flow ahead of the cooldown wave.

In addition to cooling the inner wall of the pipeline, the evaporating liquid has to absorb an increasing heat flow through the surrounding insulation as the liquid front advances. The liquid front velocity v is therefore considerably less than the figure of $S \times \rho(\text{vapour})/\rho(\text{liquid})$, i.e., less than S/1000, where $\rho(\text{vapour})$ and $\rho(\text{liquid})$ are the vapour density at ambient temperature, and liquid density respectively.

For methane at 300 K, S is about 430 m/s; hence v is less than the order of 0.4 m/s, reducing significantly as the liquid front progresses down the pipeline.

It should therefore not be surprising that it can take a long time to pre-cool a long pipeline before liquid can be transferred. Increasing the inlet pressure will not increase the mass flow rate or speed up the rate of cooldown; the exit shock will just get stronger.

It follows from these points that all transfer lines should be as short as possible.

8.9   Cooldown of a Tank with Minimum Loss of Liquid

To cool a tank down to liquid temperature efficiently, full use should be made of the "cold" in the cryogen, including both the latent heat of evaporation and the sensible heat of the vapour between boiling point and ambient temperature to cool the system from the bottom upwards. This can only be achieved if:

(1) the liquid is fed into the very bottom of the vessel, because the evaporated vapour will only flow upwards by its buoyancy inside the vessel,
(2) the vapour passes out of a vent exhausting from the top of the vessel, and
(3) the liquid is fed in slowly under low pressure difference.

The heat transfer process between cold vapour and tank wall is by natural convection, in which the heat transfer coefficient is (a) small and (b) almost independent of vapour velocity. The rate of cooldown cannot therefore be accelerated by having a high velocity vapour stream.

In other words, the cooldown is a leisurely process which cannot be hurried—if you try to hurry by increasing the flow rate, liquid will only be wasted. A sound indication that you are trying to cool down too quickly is the build-up of frost on the vent pipe.

I have demonstrated liquid cooldowns to my students on many occasions with **no** frost on the vapour vent pipe, by going easy on the cooldown flow rate and using a small pressure differential—and cooling down just as quickly as with transfers under high pressure differential with frost over all the pipework and using about 5 times as much liquid.

The secret to achieving an efficient cooldown, with minimum use of cryogenic liquid, is to let nature and natural convection take their course and then there is no stress!!

8.10   Cooldown of a Large Tank

A large tank may be provided with an overhead liquid spray near the top as well as a bottom liquid inlet. The overhead spray should not be used for cooling down the empty, warm tank since the liquid spray droplets will be carried out via the vent and their cooling capacity will be lost. The spray is intended for mixing a fresh supply with an older heel of liquid already in the tank and preventing stratification between the two liquids.

The most efficient cooldown is ensured by admitting liquid via the bottom fill inlet only.

8.11   Flashing Losses due to Transfer at Unnecessarily High Pressures

Any pressure reduction at constant enthalpy during transfer of saturated liquid will result in significant loss of liquid by flashing. For example, if LCH4 at 120 K, 3 bar (45 psi) is reduced to 112 K, 1 bar (15 psi) by expansion through a control valve, then 12 % of the liquid is evaporated and lost as a flashing loss during the transfer.

Customers object to losing 12 % of the liquid they are buying because the tanker is delivering at an unnecessarily high pressure. Two bar should be an adequate tanker pressure, causing the loss from flashing to 1 bar to be reduced to less than 6 %.

8.12   Zero Delivery

As the liquid front at the cold end of the cooldown wave advances along the delivery line, the evaporating liquid has to absorb an increasing heat flow through the insulation and therefore has a reducing capacity for cooling the pipeline. The velocity of the liquid front therefore slows down. In the extreme case of poor insulation, the liquid front may become stationary, and there cannot be any delivery of liquid.

Zero delivery can sometimes be countered by the use of cooldown vents along the line which are closed in turn as the liquid passes them.

Zero delivery can, of course, be avoided by keeping the lines short, and having adequate insulation.

8.13   Pressure Surges and the Need for Ten Second Opening and Closing Times for Liquid Valves

The sudden introduction of LNG into a warm transfer line, by the fast opening of the inlet valve, will produce a pressure surge which may lead to disaster. An initial slug of liquid will pass down the line with its front face evaporating rapidly and producing a large quantity of vapour. The pressure rise associated with this evaporation may exceed the inlet pressure and cause the unevaporated slug of liquid to flow backwards through the inlet valve. This reverse flow will be even more violent if the evaporating liquid in the hot line boils explosively by homogeneous nucleate boiling in the large temperature difference.

The reverse flow of liquid may therefore give rise to a high velocity and lead to:

- reverse liquid flow through the rotating pump, causing mechanical damage such as stripping the rotor and guide vanes,
- gate valves being slammed shut,
- line breakage and spillage.

All these have happened in the past, including a LOX transfer accident event at NASA, Cape Canaveral, USA [5].

From subsequent studies at NBS, Boulder, Colorado, it was recommended that liquid valves, whether operated hydraulically, electromagnetically or manually, should open and close over a period of 10 s or so, to avoid pressure surges and oscillations, and consequent flow reversal during liquid transfer; also that no gate valves should be used in liquid transfer lines.

It is usual for electromagnetically or hydraulically driven valves to be opened quickly in about 1 s. This is not acceptable with LNG, and a 10 s opening and closing time should be built in.

Emergency stop and vent valves should also act slowly.

8.14   Care with Topping-Out

With LNG, the boil-off vapour is usually returned or collected via a vapour return line. If the tank or vessel is over-filled when topping-out, due, say, to malfunction of the high liquid-level gauge, then liquid will pass into the warm return line, where it will evaporate and create an overpressure problem.

Great care is needed to handle this problem on the spot, remembering that:

- cryogenic liquids are compressible, and will expand with decreasing pressure,
- the tank may be brim full of liquid.

The overpressure needs to be very gently released, either by venting to the atmosphere if safe to do so, or by cracking open the valve to the boil-off gas compressor WITHOUT causing further liquid from the 100 % full tank to pass into the return line. Any panic opening of valves to the compressor will certainly pull further liquid into the return line and accentuate the overpressure build-up.

One reported topping-out incident of this nature led to a 125,000 m^3 LNG marine tanker being grossly overpressurised—and put out of commission—a very expensive mistake by the loading operator. The tanker had to be emptied at the loading terminal, dry-docked and one or more inner membrane tanks removed and rebuilt! In addition, some LNG spilled on to the mild steel deck and caused the plates to brittle fracture.

It is concluded that topping out a 1 bar storage or transport tank should be to less than 99 % full, not 100 %.

References

1. Martinelli, R.C., Lockhart, R.W.: Proposed correlation of states for isothermal 2-phase flow in pipes. Chem. Eng. Prog. **45**(1), 39 (1949)
2. Richards, R.J., Steward, W.G., Jacobs, R.B.: Transfer of liquid hydrogen through uninsulated lines. Adv. Cryog. Eng. **5**, 103 (1960)
3. Shen, P.S., Jao, Y.W.: Pressure drop of 2-phase flows in a pipeline with longitudinal variations in heat flux. Adv. Cryog. Eng. **15**, 378 (1970)
4. Shapiro, A.H.: The Dynamics and Thermodynamics of Compressible Fluid Flow. Ronald, New York (1953)
5. Edeskuty, F.J., Stewart, W.F.: Safety with Handling of Cryogenic Fluids. Plenum, New York (1996)

Chapter 9
Safe Handling and Storage of LNG and LPG

Abstract Every year, people have died from accidents with cryogenic liquids. The enormous volume expansion of 600 to one for LNG, and 350 to one for LPG, takes everyone by surprise when there is a leakage or accidental spill. The vapours are non-toxic and odourless, and can easily reduce the oxygen level below that needed for respiration. There is no physiological warning of oxygen deficiency, and asphyxia will kill very quickly.

If handling a cryogenic liquid in a confined space, beware of the danger of asphyxia: carry a reliable oxygen meter to provide continuous monitoring of the air you are breathing.

This chapter includes a number of advisory points on safety by the author, from some 60 years of storing, handling and carrying out research on all kinds of cryogenic liquids, including LNGs, LPGs and Freon mixtures.

9.1 Summary of Safety Points Raised

1. Read the BCC *Cryogenics Safety Manual: A Guide to Good Practice* for a more comprehensive guide to safety.
2. Be aware of the deadly danger of asphyxia, especially when the gases are non-toxic and give no physiological warning of oxygen deficiency, like nitrogen, and hydrocarbons.
3. Do not overfill vessels.
4. Beware of stratification, and the consequences.
5. LPG spills are particularly hazardous because they produce heavy flammable vapour clouds which disperse very slowly into the air.

9.2 General Remarks

The need to conduct the handling and storage of cryogenic liquids in a safe and responsible manner is obvious for moral, environmental and economic reasons.

© The Author(s) 2016   121
R.G. Scurlock, *Stratification, Rollover and Handling of LNG,*
LPG and Other Cryogenic Liquid Mixtures, SpringerBriefs in Energy,
DOI 10.1007/978-3-319-20696-7_9

A good first point of reference is "*The Cryogenics Safety Manual*: *A guide to good practice*" published by the British Cryogenics Council [1].

Safety has always been a basic consideration for anyone working with cryogenic materials and more recently this has been covered by regulations. Specific safety guidelines have been laid down by national and international legislation as in the US and Europe. In the UK, the Health and Safety Executive (HSE) is the working body for guidelines on safety procedures, accident reporting and analysis, and for legal action if necessary.

The Cryogenics Safety Manual is a useful guide to good practice for all operators and users handling cryogenic fluids. The Manual was first published in 1970 and it has since been revised in 1982, 1991 and 1998 with a fifth edition planned for 2016.

Other references on safety include the volume by Zabetakis, on *Safety with Cryogenic Fluids*, published in 1967 [2], and the volume by Edeskuty and Stewart on *Safety in the Handling of Cryogenic Fluids*, published in 1996 [3].

This chapter does not intend to cover all the safety material in the Cryogenics Safety Manual. It does, however, contain a useful description of the building and operation of a cryogenic safety laboratory. It also contains many advisory points, arising from earlier chapters of this book, which are additional to those made in the *Cryogenics Safety Manual*, and includes some personal experiences of handling many cryogenic liquids over some 60 years.

9.3 Health Concerns

9.3.1 Cold Burns

Human tissue cooled below about −10 °C will suffer necrosis or frost bite or cold burns. It is therefore sensible to ensure that, when handling cryogenic liquids, gases and cold equipment, the right clothes are worn so as to avoid cold burns. These should include loose gloves, long sleeves, trousers without turn-ups, shoes/boots—not sandals—eye protectors and so on. First Aid for a cold burn is the application of cold water (*not* hot water) as fast as possible, since seconds wasted will make all the difference between serious injury and just a stinging sore patch.

9.3.2 Asphyxia and Anoxia

Any cooldown process, or spillage of cryogenic liquid, will result in the generation of large volumes of cold, dense vapour. Cold vapour will firstly flow downwards and displace atmospheric air upwards, and secondly mix with air, diluting the oxygen content and forming a dense cloud of condensed water particles. The cloud, and the clear space below it down to ground level, is deficient in oxygen and is therefore dangerous to breathe. If the oxygen content is less than 6 % then sudden and acute asphyxia will result in instant unconsciousness, without any physiological warning. As described in the BCC Cryogenic Safety Manual, "The victim will fall to the ground as if struck down by a blow to the head and will die in a few minutes unless immediate remedial action is taken".

Fatal accidents happen every year from sudden asphyxia. If a major spillage occurs, evacuate the area immediately, and keep away from the cloud of cold vapour.

Gradual asphyxia, or anoxia, also with fatal consequences, can arise from progressive lowering of the oxygen content by poor venting procedures over a period of time. Asphyxia can be avoided by the use of firstly, common sense and secondly, personal gas monitors, which produce an alarm signal if the oxygen content drops below 20.9 % (say to 19 %). The important thing is to be aware of the deadly danger of asphyxia, with no physiological warning of oxygen deficiency (breathlessness is only brought on by the build up of carbon dioxide) and the danger of complacency about the hazards leading to asphyxia.

9.4 Equipment Failure

9.4.1 Materials

Certain materials are not suitable for use in cryogenic systems because they suffer a ductile to brittle transition as the temperature is reduced below ambient, which can lead to brittle failure. Such materials include, for example, carbon steels and plastics and they must not be used.

Generally, metal fatigue limits are higher at low temperatures so that fatigue failure is less likely. However, brittle failure of pipes and tanks can lead to spillages and serious consequences.

9.4.2 Overpressure

Overpressure in a pipe line, following shutdown after a transfer, must be prevented by protecting each valved-off section with its own pressure-relief valve and adequately sized vent. Vacuum spaces are a particular problem, because a leak from the inner liquid-containing vessel can lead to overpressure in the vacuum space and subsequent collapse of the inner vessel. All vacuum spaces should be protected by blow-off discs to protect against overpressure.

9.4.3   Spillage Containment

A major spillage of cryogenic liquid can lead to serious and tragic consequences, so the concept of secondary containment of the spill for a short period needs to be applied. This can take the form of:

(1) A berm, bund or earth bank to prevent a major spill spreading into the local environment.
(2) A secondary, liquid-containing wall, as for LNG tanks.
(3) A secondary barrier or membrane within the insulation space, as in LNG ship tankers, to prevent contact between LNG and the carbon steel structure of the ships. Additionally, it has been found that the secondary membrane also provides considerable anti-spill protection during ship-ship collisions and ship-shore groundings.
(4) Adequate integrity of the outer containment structure of the applied insulation, as for VI pressure tanks.

The nightmare scenarios of spillage accidents generally arise with single skinned containers such as LPG pressurised road and rail tankers and the older crude petroleum tankers.

Spillage due to transfer line breakages can be prevented by the installation of automatic shut-off valves inside storage tanks and vessels.

9.4.4   Fire and Explosion

The conventional description of fire is the coincidence between:

(1) combustible
(2) oxidant and
(3) ignition source.

Fire protection is the rigorous separation of these three items; a combination of any two represents a potential fire hazard.

Hydrocarbons, such as LNG or LPG, are generally flammable in air over limited composition ranges. For example, methane is only flammable with compositions from 5 and 15 % by volume of methane in air. The density of methane gas is about one half that of air, so the vapour from LNG rises rapidly, mixes by convection with the air and quickly dilutes to below the lower flammability limit of 5 % while dispersing upwards and away from ignition sources.

Heavier hydrocarbons, like LPG, which evaporate to produce a vapour more dense than air present a more serious fire hazard. The vapour stratifies and (a) does **not** mix convectively with less dense air, and thereby tends not to dilute to below the lower flammability limit, and (b) does **not** disperse by rising upwards.

Apart from ignition sources created by high or low voltage electrical faults, it should be remembered that electrostatic charges can also create ignition sources in the form of sparks. All cryogenic liquids have exceedingly low electrical conductivities and their movement from vessels and tanks through pipelines or hoses, which are not adequately earthed, can induce charge separation and high voltages.

9.5 Liquid Management

9.5.1 Overfilling of Vessel or Tank

Unlike water, LNG and LPG have finite volume coefficients of compressibility, and they expand significantly with increasing temperature along the saturation line (see Table 5.1), or with isenthalpic and isentropic decreases in pressure. This behaviour is totally unlike that of water, which has almost zero compressibility; thus water is a useful liquid, for hydraulic power transmission and pressure testing of vessels.

If a vessel is overfilled with LNG, then the additional boil-off vapour escaping through the vent-line will create a back-pressure, compressing the liquid. If the additional boil-off is now relieved by opening a second vent or pressure relief valve, the pressure falls, the liquid may expand from the already full vessel into the vent-line, and unexpectedly increases the boil-off mass flow, causing the vent-line pressure to rise still further. What action should be taken?

The first thing to realise is that the problems of overfilling a vessel do not stop when the LNG transfer is cut off. The compressibility of the liquid must be anticipated by slowly opening additional vapour vents, without encouraging liquid to expand into the vent-line.

Secondly, if LNG spills, or sprays, out of the vents, beware of (a) cold burns to personnel, (b) asphyxiation by displacement of breathing air by dense cold vapour and (c) cold damage to carbon steel structures and wet concrete pavings.

It is therefore important to ensure that emergency procedures are in place that permit the immediate and controlled evacuation of personnel to avoid (a) and (b), and the setting up of water sprays to avoid (c).

Clearly, upper liquid level alarms should be accurate to better than 1 % and be set to allow sufficient time for corrective action, before overfilling occurs.

9.6 Safety Laboratory Features

To carry out tests, demonstrations and research programmes on flammable liquids, a safety laboratory is required. Such a safety laboratory was built at the Institute of Cryogenics, Southampton University, which has enabled research work to be carried out continuously and safely, with liquid hydrocarbons, for many years. With the increase in safety regulations being imposed, the cryogenic safety laboratory was an area of specific concern for Health and Safety inspectors. The following features were therefore installed to meet their safety requirements or cryogenic liquid hydrocarbons and liquid hydrogen.

9.6.1 Fire and Explosion Containment

The obvious answer was a separate and extensive purpose-built building, but instead an existing single storey laboratory, already integrated into a larger research building, was adapted to enable the use of existing common services and utilities.

The walls of the laboratory were made fire resistant so as to contain a conflagration for 30 min. The roof of the single storey laboratory was soft enough to blow off if an explosion occurred, without damaging the wall structure. The severity of any possible fire/explosion was controlled by an edict limiting the total amount of flammable liquid contained in vessels within the laboratory. In the event of a fire and evacuation of the laboratory, the ventilation fans and vents could all be turned off and closed from outside the laboratory, so as to restrict incoming oxygen, and aid extinguishing the fire by oxygen starvation.

9.6.2 Ventilation

Two large fans were installed in the outside wall, one at high level to deal with methane, one at low level to deal with higher hydrocarbons. The fans were large enough to ensure a complete change of air in the room every 20 s. Large one-way vents, on the entrance door in the opposite wall, ensured a cross-flow of air from inside the building. The fans were driven by induction motors which were fully rated for operation in a flammable atmosphere.

In practice, the laboratory air was monitored by eight semi-conductor gas detectors placed at high, low and bench levels around the laboratory, all connected to a switchable indicator and common audio alarm. As a result, the fans could be throttled back or turned off, as necessary, to make the laboratory a more comfortable working environment over long periods.

9.6.3 Personnel Safety

The number of personnel in the laboratory at any one time was limited, so that evacuation through the entrance door or the emergency exit, to the outside of the building, could be achieved within seconds.

The floor was surfaced with a proprietary carbon-loaded, electrically conducting medium and all personnel had to wear earthing strips on their shoes so as to reduce electrostatic sparking from clothes of synthetic materials.

Generally, students were encouraged to wear cotton-based, rather than synthetic, top clothes and underwear when working in the safety laboratory, to avoid electrostatic sparking within their clothes.

9.6.4 Management of Research Experiments

The laboratory bench area was covered by a large hood connected to a third exhaust fan so that experimentally generated flammable gas from open rigs was vented separately out of the laboratory.

The fan exits through the roof had to be positioned away from the edges so as to reduce the likelihood of a lightning strike during a thunderstorm.

An additional multi-point access vent line, positioned all round the laboratory and exhausting into the fan entrance was used to take boil-off gas from experiments and storage vessels.

These arrangements enabled liquid methane, LNG and liquefied ethane, propane and butane to be handled in both open-top and closed dewars, cryostats and containers, just like LIN.

In this way, students were able to gain hands-on experience, and meet at first hand the limits of measurement with instrumentation, caused by some of the peculiar properties of LNG, with its foaming, spitting and Marangoni wall films.

9.6.5 Asphyxiation and Toxic Gases

The use of portable, battery operated, oxygen composition meters were encouraged at all times, when using cryogenic liquids in closed areas. The oxygen meters all had audio alarms set to signal if the oxygen levels fell from the normal 21 to 19 %. They could be tested quite simply by blowing on the detector port—exhaled breath has an oxygen level below 19 % and is guaranteed to sound the alarm.

The venting procedures outlined in Sect. 9.6.2 prevented oxygen levels falling below 20 % in the safety laboratory, so that the risk of asphyxiation was minimised. Hydrocarbon gases are very slightly toxic and only then at high levels of contamination, the occurrence of which was prevented by the venting system. Some students acquired headaches after working with propane and butane in the laboratory for periods of many hours.

The main problems were caused by impurities and additives. For example, natural gas has mercaptans added as an odoriser, and we found that liquefaction of natural gas from the mains concentrated the odoriser in the condenser. A single whiff made one feel very sick for the rest of the day.

9.7 Particular Liquids

The following are personal comments on particular cryogens, based on the author's continuous working experience since 1954. They should be regarded as additional to the standard advice in safety literature.

9.7.1   Methane 112.2 K, Ethane 184.2 K, Ethylene 169.2 K, Propane 231 K and n-Butane 272.6 K

As single component liquids, boiling below ambient, all these hydrocarbons have liquid handling properties like those of LIN; with the flammability of their vapours as an additional property. In a fire situation, it is the vapour which burns and so the size of the flame above an open burning pool of liquid is directly related to the evaporation rate. A short time after a spillage on to the ground, the heat flux through the frozen sheath of earth falls significantly and it is only the surface heat flows from the atmosphere and the radiation from the flame which contribute to the liquid evaporation.

Thus, covering the surface of the liquid pool with a foam blanket, to separate liquid and vapour, provides insulation from these two heat flows and an immediate reduction in boil-off rate, thereby smothering the flame.

9.7.2   Liquid Natural Gases LNGs

LNG is a mixture, and exhibits some peculiar properties, which make it rather unpleasant to handle, and difficult to obtain accurate in-tank instrumentation readings, in comparison with, say, LIN.

Firstly, as a mixture, it exhibits the Marangoni effect, whereby the walls of the containing vessel, and instruments above the liquid surface, are covered in wet fluid and droplets of ethane-rich liquid.

Secondly, it spits and splashes. Looking into an open, insulated bucket of LNG is hazardous because any motion of the low viscosity liquid, produced by touching or moving the bucket, will result in spitting, i.e., a series of small vapour explosions at the container wall and the ejection of many droplets of liquid into your face, or up to the ceiling if they miss you. This spitting appears to be a property of LNG and not of LCH4. I understand that this spitting from sloshing can be heard on LNG tankers in a heavy sea and the associated local vapour explosions may cause damage to the walls of the membrane tanks.

Thirdly, spillage of LNG on to water is also accompanied by irregular explosive boiling. This phenomenon, called Rapid Phase Transfer or RPT, is alarming, but is probably not dangerous in the open air [4]. The evaporated vapour is generally non-polluting and buoyant because it has a density less than that of atmospheric air above about 170 K (−150 °C), becoming about one half the density of air at ambient temperature. The vapour therefore mixes almost spontaneously with air to dilute itself below 5 %, the lower limit of flammability, and rises rapidly away from possible ignition sources. LNG release should be avoided as far as practical due to the "global warming" potential of Methane gas.

On the other hand, spillage into underground tunnels and tunnel networks, like sewage systems, is extremely hazardous and must never be allowed to happen.

The vapour is unlikely to be diluted with air to below 5 % in the tunnel and a flammable or explosive mixture may be formed. It only requires an ignition source, such as an electric spark, to initiate an underground explosion and widespread damage above ground.

9.7.3 Liquefied Petroleum Gases (LPGs)

Liquefied Petroleum Gases are largely mixtures of propane with normal- and iso-butane. Large quantities are produced by petroleum refineries as non-condensable gases, which used to be regarded as waste gases and were burnt in flare stacks. But not today because they are a valuable source of energy and now form part of industrial cryogenic activity. The storage, handling and distribution of LPGs in large quantities, as refrigerated liquids down to −42 °C at 1 bar, or as ambient temperature liquids under pressures up to 30 bar, is probably more widespread than the LNG industry.

LPGs do however present a safety hazard, because the vapours are more dense than air by a factor of about two. This means that any spillage, of liquid or vapour, is accompanied by the production of a low-lying cloud of heavy flammable vapour close to the ground or water surface.

Unlike natural gas, this LPG cloud does not spontaneously mix with air and dilute itself below the lower limit of flammability, and can flow over the ground, guided by topographical features for a long distance. Most ignition sources are at ground level, so the situation is easily reached for a major fire to occur from a relatively small spillage.

If a fire does commence, then the possible disastrous behaviour of a single or double skinned storage tank of LPG, whether refrigerated or pressurised, must be anticipated and evacuation of the immediate area to a distance of several hundred meters is essential. This is because several LPG tanks have been known to disintegrate in a fire via a Boiling Liquid Expanding Vapour Explosion, or BLEVE, creating a fireball several hundred meters in diameter, and causing many casualties.

Because of the growing scale of the LPG industry, very large tanks of 100 m diameter have been constructed for storing refrigerated liquid down to −42 °C. The tanks are constructed of pre-stressed concrete, including the roofs, which are designed to withstand limited storage pressure excursions above 1 bar, measured in inches water gauge.

Since LPGs are mixtures, they suffer from all the evaporation instability and path-dependent mixing behaviours met with LNG. Thus, provision must be made for stratification and its removal by adequate mixing, and the possible occurrence of rollover and vapour explosions. More importantly, there must be adequate emergency venting to release associated BOR spikes and thereby avoid over-stressing the concrete structure of the tank, which may not be designed to withstand the pressures associated with boil-off instabilities.

References

1. The Cryogenics Safety Manual; A Guide to Good Practice. British Cryogenics Council, 5th edn (2005/6)
2. Zabetakis, M.G.: Safety with Cryogenic Fluids. Plenum, New York (1967)
3. Edeskuty, F.J., Stewart, W.: Safety in the Handling of Cryogenic Fluids. Plenum, New York (1996)

Index

© The Author(s) 2016
R.G. Scurlock, *Stratification, Rollover and Handling of LNG,*
LPG and Other Cryogenic Liquid Mixtures, SpringerBriefs in Energy,
DOI 10.1007/978-3-319-20696-7

Printed in the United States
By Bookmasters